Thinking Spatially Using GIS

ISBN-13: 9781589481800
Suggested
grade level: 3–6

Thinking Spatially Using GIS presents engaging lessons that use fundamental spatial concepts and introduce GIS software to young students. Students are guided to find relative and absolute locations of map features, create maps, locate human and physical features on maps, discover and analyze geographic distribution patterns, and investigate changes over time. *Thinking Spatially Using GIS* provides a platform to tie in conventional 3–6 grade level topics, including lessons learned on world and U.S. geography, world exploration, and demographics.

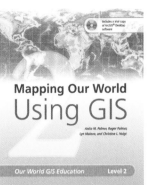

Mapping Our World Using GIS

ISBN-13: 9781589481817
Suggested
grade level: 6 and up

Mapping Our World Using GIS invites middle and high school students to investigate GIS through lessons that require critical thinking and problem-solving skills. These lessons align with national teaching standards for geography, science, and technology in introducing and building skills in geographic inquiry, spatial thinking, and GIS. Topics covered include studying landforms, physical processes, ecosystems, climate, vegetation, population patterns and processes, human and political geography, and human and environment interaction.

Analyzing Our World Using GIS

ISBN-13: 9781589481824
Suggested
grade level: 9 and up

In *Analyzing Our World Using GIS,* students gain proficiency working with GIS and exploring geographic data. High school and college students will complete sophisticated workflows such as downloading and editing data and analyzing patterns on maps. Topics covered include analyzing economics by exploring education funding, demographics, and trade alliances; analyzing land and ocean effects on climate; and plate tectonics analysis to describe earthquake and volcanic activity.

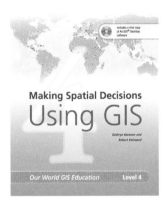

Making Spatial Decisions Using GIS

ISBN-13: 9781589481831
Suggested
grade level: 13 and up

Making Spatial Decisions Using GIS is intended for advanced high school, college, university, and technical school students. The book presents a wide variety of real-world settings for GIS analysis and decision making. Students learn methods for planning and executing GIS projects for more involved group investigations and independent study. Topics covered include analyzing hazardous emergency situations, using demographic data for analysis of population growth and urbanization trends, and using GIS to create reliable location intelligence that results in sound decisions.

Thinking Spatially
Using GIS

Eileen J. Napoleon

Erin A. Brook

ESRI PRESS

REDLANDS, CALIFORNIA

Ask for ESRI Press titles at your local bookstore or order by calling 1-800-447-9778. You can also shop online at www.esri.com/esripress. Outside the United States, contact your local ESRI distributor.

ESRI Press titles are distributed to the trade by the following:

In North America:
Ingram Publisher Services
Toll-free telephone: (800) 648-3104
Toll-free fax: (800) 838-1149
E-mail: customerservice@ingrampublisherservices.com

In the United Kingdom, Europe, and the Middle East:
Transatlantic Publishers Group Ltd.
Telephone: 44 20 7373 2515
Fax: 44 20 7244 1018
E-mail: richard@tpgltd.co.uk

Cover design and production   Jennifer Campbell
Interior design   Jennifer Campbell
Interior production   Savitri Brant
Image editing   Jay Loteria
Editing   Mark Henry
Copyediting and proofreading   Tiffany Wilkerson
Permissions   Kathleen Morgan
Printing coordination   Cliff Crabbe and Lilia Arias

On the cover
World map by Michael Law.
Cover data from ESRI Data & Maps 2006, courtesy of ArcWorld Supplement and National Geophysical Data Center.

Contents

Preface

We have written this book to introduce the basic concepts of geographic information systems (GIS) and geography in a classroom setting. The lessons in this book will help build a bridge between the classroom learning experience and the world outside. From these lessons, students in the elementary grades and beyond will gain hands-on experience using GIS technology as a tool, applying it to a broad range of topics, with a special focus on geography.

Our World GIS Education, Level 1: Thinking Spatially Using GIS puts actual spatial data into the hands of students to visualize, analyze, interpret, and study various scenarios, including exploration of the New World, tornado patterns in the Great Plains, America's westward migration, and even a visit to the local zoo. Displaying this data visually will help students understand how GIS can help them make decisions that affect their families, classrooms, communities, nation, and the world.

The lessons in this book will help teachers lead students through exercises using GIS maps and will also build skills in critical thinking and decision making. We have chosen relevant topics that students are likely to encounter in their classroom studies and in their daily lives. As students do these lessons, they also can have fun finding solutions to problems faced by everyone from ancient explorers to modern-day tornado chasers.

This book comes with a CD. It contains the GIS data and other documents students will need to complete the lessons, using the free software, ArcExplorer Java Edition for Education (AEJEE), for Macintosh or Windows computers. This book also provides information about online resources, and there is a companion Web site, www.esri.com/ourworldgiseducation.

Thinking Spatially Using GIS is the first volume in *Our World GIS Education*, a new four-book series of comprehensive GIS instruction for students of all ages, from elementary school level to college undergraduates. The series builds on the solid foundation of *Mapping Our World: GIS Lessons for Educators*, the popular ESRI Press book geared to middle and high school classrooms.

Thinking Spatially Using GIS presumes no previous experience or skill in using geographic information systems. This beginning text is appropriate for anyone who wants to explore and learn about how this powerful technology works in the real world. This book includes a section on how to use the material, and each of the lessons includes teacher's notes to explain the student activities. We have provided student activities, worksheets, and answer keys for your convenience.

We hope that the teachers who present these lessons and the students who complete them will enjoy learning how to think spatially using GIS!

Eileen Napoleon and Erin Brook

About the authors

Eileen Napoleon has worked as a geographer, first for the U.S. Forest Service, and for the past 15 years as a GIS education specialist at ESRI in Redlands. Eileen is still passionate about geography and education, and especially enjoys teaching GIS and geographic concepts to those who are new to the subject.

Erin Brook spent five years working for ESRI Canada as the K-12 industry manager. She was involved with the marketing and development of ESRI's education programs for K-12 students. She incorporated GIS directly into required geography curriculum and made GIS technology available to students across Canada. Erin now works as an education specialist for ESRI in Redlands, California.

Acknowledgments

We would like to thank all those who helped make this book possible:

At ESRI: Laura Bowden, who managed this project; Joseph Kerski and Charlie Fitzpatrick, who reviewed all of the exercises and provided valuable input; David Davis, for refining some of the data; and Brandy Perkins, who tested the book's exercises. Nick Frunzi, ESRI Educational Services director, and Jack Dangermond, ESRI founder and president, deserve special thanks for their vision and support of this project.

At ESRI Press: Mark Henry, who edited the content; Jennifer Jennings, who designed the book; Michael Law, who refined the cartography; Kathleen Morgan, who oversaw permissions; Tiffany Wilkerson, who copyedited the manuscript; Savitri Brant, who designed the interior of the book; and Jay Loteria, who handled digital content. Judy Hawkins, former ESRI Press manager, and her successor, Peter Adams, provided outstanding leadership.

We also want to thank Kathryn Keranen, Bob Kolvoord, Anita Palmer, and Roger Palmer for reviewing lessons and providing feedback and direction early in the project. We are especially grateful to Joan Smith, elementary science teacher for the Rainier School District; Mark Lowry, Geography and Geotechnologies consultant for the Toronto District School Board; and Dr. Alan Richard Buss, head of the Department of Elementary and Early Childhood Education for the University of Wyoming, for reviewing the book and providing expert guidance.

We also want to acknowledge the wonderful students who participated in this project at the ESRI Annual User Conference Kids Camp.

Introduction

About the lessons

Thinking Spatially Using GIS is a book of computer activities, data, and resources that can help introduce students at young ages to mapping concepts, GIS, geography, and other relevant topics. Because the lessons cover topics that you are already teaching in social studies, geography, history, life science, and earth science, we think you'll find it to be a valuable supplement to your current textbook or curriculum. We've designed these lessons to help your students get excited about exploring the world around them while stepping into the world of contemporary computing and the smart maps of geographic information systems.

The book comes with a copy of ArcExplorer Java Edition for Education (AEJEE) GIS software. Students will use this software to work through the lessons. This means that your students will be learning to use GIS tools that are similar to those being used by professionals around the world. And they will be using tools that they are increasingly likely to encounter in the future, whether it is in middle or high school, college or technical school, or as a citizen on the Internet.

Where to begin

Before you start using the book in class, we recommend that you do the following:

1. Finish reading this section and skim through this book and the student workbook to locate teacher and student materials.
2. Install the teacher resources, AEJEE software, and lesson data on your computer. (Besides being there for your own use, this can serve as a backup of the lesson data and teacher resources in case the disk is misplaced.)
3. Work through all of the lessons yourself.
4. Choose the lessons that you would like to present to your students.
5. Read the "Teaching the lesson" section of the teacher materials for ideas on how to present the lessons you choose.
6. Install the AEJEE software and lesson data on student computers. (Refer to "Setting up the software and data" later in the next section, and the detailed software and data installation guide at the back of the book.)
7. Work through the lessons of choice with your students.

Modules

The modules in this book present spatial thinking concepts. Each module has a theme on spatial thinking. In module 1, the theme is location—of places (absolute and relative) and the characteristics of locations. In module 2, the theme is classification—making sense of the world by organizing the things in it into categories and quantities. In module 3, the theme is movement and patterns —the migration of humans and the patterns that result. In module 4, the theme is analyzing change—over space (spatial analysis) and over time.

While all of the modules are designed to be appropriate for elementary school, some are more appropriate for certain grades.

Module	Appropriate grade levels
1	3rd through 6th grade
2	4th through 6th grade
3	5th through 6th grade
4	4th through 6th grade

You can teach each module independently of the others, and you should tailor the material to suit the specific needs of your class and curriculum. Each module consists of two or three lessons. Within some modules, the lessons build on each other. This is true in modules 1 and 2, while the lessons in modules 3 and 4 can be presented independently of one another.

Module	Recommended order
1	Lessons 1 and 2 should be presented in order.
2	Lessons 1, 2, and 3 should be presented in order.
3	Lessons 1 and 2 can be presented independently.
4	Lessons 1 and 2 can be presented independently.

You may also choose to present only one lesson within a module.

Prerequisite skills

Students are expected to have basic computer skills such as using a mouse, highlighting text, and moving and closing windows. For a guide on preparing students to work with AEJEE software, refer to "Preparing students to work with AEJEE software" in the next section.

Prerequisite knowledge

Students will get more out of the lessons if they have some of the prerequisite knowledge. The "Introducing the lesson" section of the teacher materials for each lesson has suggestions on what to review before students work through the GIS activities. For example, before working through the first GIS activity in module 1, students would benefit from understanding the major lines of latitude and longitude (e.g., equator, prime meridian, Tropic of Cancer, Arctic Circle) and how they divide the world into hemispheres and zones or regions (e.g., tropics).

How the materials are organized

This book contains teacher materials for each lesson, including the following:

- A lesson overview, including estimated time and list of materials for the lesson, objectives, a list of the GIS tools and functions encountered in the activity, and a list of national geography standards related to the lesson.
- Notes about teaching the lesson, including suggestions for introducing the lesson, teacher tips on how to conduct the GIS activity, things to look for while students are working, and how to conclude the lesson. Ideas for extending the lesson are listed as well.
- A list of Web sites with information related to the lesson.
- Student activities and worksheets with answers.

The accompanying student workbook contains GIS activities, worksheets, and handouts.

Options for providing these materials to your students include ordering copies of the student workbook, photocopying the student workbook pages, or printing out the student workbook pages from the files on the CD.

The software, data, and resources CD has three separate installations: the AEJEE software, a student data folder, and a teacher resources folder.

The teacher resources folder (OurWorld1_teacher) includes the following:

- Digital documents (PDF format) of the student activities, worksheets, and handouts required by the lessons.
- Digital documents (PDF format) of some of the teacher materials in this book, such as lesson overviews, teacher notes, and student activities with answers.

The AEJEE software can be installed on Windows or Macintosh computers. (Refer to the section, "About the software and data.")

The student data folder (OurWorld1) includes the data that students will access and view with the AEJEE software. This folder needs to be installed on each student computer after the AEJEE software is installed. Inside the OurWorld1 folder, the data is organized by module.

Curriculum standards

Each lesson teaches skills corresponding to National Geography Standards (*Geography for Life: The National Geography Standards*, 1994). The applicable standards for elementary school (grades K-4) and middle school (grades 5-8) are listed in each lesson overview. Matrices matching all lessons in the book to the National Geography Standards and to the National Technology Standards are also provided (see table of contents).

The companion Web site

The book's Web site, www.esri.com/ourworldgiseducation, places a variety of GIS resources and other helpful information at your fingertips. For example, you'll want to check the Web site's "Resources by Module" section for specific resources, Web links, or changes when you get ready to use a particular lesson. Solutions to common problems and any significant changes or corrections to the book will also be posted here.

Taking it further

After your students have completed the lessons you selected, you can do the following:

- Challenge the students further in the "Extending the lesson" part of the teacher notes.
- Challenge them with another AEJEE-based lesson that you download from ArcLessons on the ESRI website: www.esri.com/arclessons.
- Have them put together a profile of your community and post it on the ESRI Community Atlas Web site: www.esri.com/communityatlas. Your school may be able to earn software through this program.
- Find out who's doing what with GIS near you and contact them for ideas. The following resources and suggestions can help:

  - ESRI GIS Education Community, http://edcommunity.esri.com
  - ESRI Education User Conference, www.esri.com/educ.
  - GIS Day Web site, www.gisday.com
  - GIS.com Web site, www.gis.com
  - Invite a GIS specialist from your city government or other local organization to do a presentation on GIS for your class.
  - Challenge 5th or 6th grade students with lessons from level 2 in the *Our World GIS Education series: Mapping Our World Using GIS*. This book comes with a 365-day trial version of ArcGIS Desktop software (ArcView license). You may want to check with your district or state technology coordinator to find out if a district-wide or statewide software license already covers your school.

About the software and data

ArcExplorer Java Edition for Education software

The lessons in this book use ArcExplorer Java Edition for Education (AEJEE) software. AEJEE is free software from ESRI that is provided on the CD at the back of this book. The software is also available for download from http://edcommunity.esri.com. AEJEE can be used to display and visualize geographic data, perform basic GIS analysis, and create basic map layouts for printing.

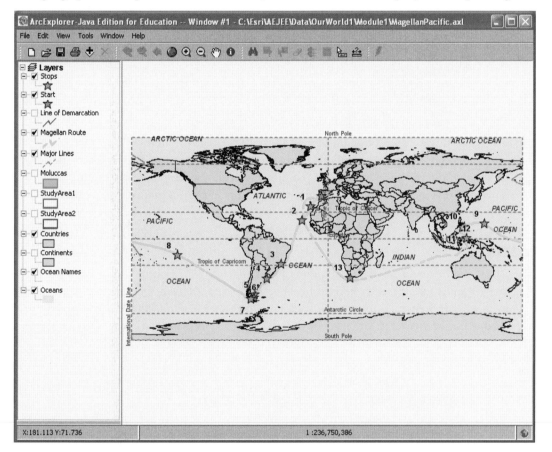

Setting up the software and data

This book comes with a CD that contains AEJEE software for Macintosh or Microsoft Windows computers, as well as the data, student activities, and other documents required for the lessons. Refer to the installation guides at the back of the book for detailed system requirements and instructions on how to install the software, data, and resources. A copy of the data license agreement is also found at the back of the book.

Note: It is important that you install the AEJEE software before installing the data, since the data will be installed in one of the AEJEE folders.

If you do not feel comfortable installing programs on your computer or your students' computers, please be sure to ask your school's technology specialist for assistance. The software and data on the disk needs to be installed on your computer and all computers that the students will use to complete the GIS activities. The teacher resources should be installed on your computer, not on the student computers.

Desktop shortcut. Consider creating an AEJEE shortcut icon on the desktop of each student's computer so that students can quickly locate and start the AEJEE program. (Refer to page 256 and 266 for instructions on how to create a shortcut on a Windows-based and Macintosh computer, respectively.)

Student work folders. The activity instructions advise students to ask their teacher how and where to save their work. It is a good idea to have students save their work to a student folder. This way the original data is unaltered and can be used by multiple students. See the "Teacher notes" section of individual lessons for more information.

Moving or deleting data. The installation program installs the data for all modules. If for some reason you decide to remove part of the data, be sure to keep entire module folders intact. For example, if you only plan to teach lesson 1 of module 2, you should keep the entire module 2 folder. Lessons within a module often share some data, and map projects are set up to find data in specific locations within a module folder.

Metadata

Metadata (information about the data) describes the GIS data provided on the Software and Data CD. The metadata includes a description of the data, where it came from, a definition for each attribute field, and other useful information. Each data file has a corresponding metadata file in text (txt) format that can be viewed in any text editor (e.g., Notepad or Wordpad on the PC). A copy of each metadata file in XML format is also included, which can be viewed in ArcGIS Desktop software, if you happen to have it.

Preparing students to work with AEJEE software

Students will get more out of the lessons if they have some basic computing skills. Here is a list of computer skills to review before students work through the GIS lessons.

- Left-clicking, right-clicking (PC only), single-clicking, and double-clicking.
- Clicking and dragging, for example, moving a window to a different location on the computer screen or zooming in using the Zoom In tool.
- Moving the mouse to position the cursor.
- Maximizing the AEJEE window.
- Highlighting text to select and replace it.
- Navigating to different data folders and selecting files. For example, to open a map (AXL file) for the lesson, students must navigate to the proper module folder. To add a layer (shapefile) to the map, students must navigate to the proper data folder.

Note: Exercise instructions are written for Microsoft Windows users. Students working on Macintosh computers should review the Macintosh user guide at the back of the book.

Here are general guidelines and hints for helping students work through the GIS lessons. See the "Teacher notes" section of individual lessons for more information.

- For younger grades, lead students through the lessons. For example, perform one or more GIS actions while the students follow along, repeating your actions. This is effective for students with limited reading skills.
- Allow one or more hours for each lesson.
- If you plan to complete more than one lesson, book the computer lab for 2 or 3 days in a row if possible.
- Modify the lessons to include your own questions and steps.
- Data is installed locally on each student's computer. If you are using the lessons with more than one group of students, you should create student work folders and have students save their work to these folders.

Troubleshooting ArcExplorer

Once in a while, the software may not perform as expected. Often, simple solutions correct the situation. For example, if a map layer does not display properly when you turn it on, turn the layer off and then turn it on again. This usually resolves the issue. A list of commonly encountered troubles and their solutions can be found on this book's Web site, www.esri.com/ourworldgiseducation.

If you have questions related to installing the software or you want to report a problem or error with the lesson materials in this book, you can send e-mail to learngis@esri.com with your questions.

Correlation to National Geography Standards

Standard	Module 1		Module 2			Module 3		Module 4	
	Lesson 1	Lesson 2	Lesson 1	Lesson 2	Lesson 3	Lesson 1	Lesson 2	Lesson 1	Lesson 2
1. How to use maps and other geographic representations, tools, and technologies to acquire, process, and report information from a spatial perspective	✓	✓	✓	✓					
2. How to use mental maps to organize information about people, places, and environments in a spatial context	✓	✓	✓	✓					
3. How to analyze the spatial organization of people, places, and environments on Earth's surface	✓	✓					✓		
4. The physical and human characteristics of places									
5. That people create regions to interpret Earth's complexity					✓			✓	✓
6. How culture and experience influence people's perceptions of places and regions									
7. The physical processes that shape the patterns of Earth's surface								✓	✓
8. The characteristics and spatial distribution of ecosystems on Earth's surface					✓				

Source: Geography Education Standards Projects. 1994. *Geography for Life: National Geography Standards 1994.*
Washington, D.C.: National Geographic Research and Exploration.

Standard	Module 1		Module 2			Module 3		Module 4	
	Lesson 1	Lesson 2	Lesson 1	Lesson 2	Lesson 3	Lesson 1	Lesson 2	Lesson 1	Lesson 2
9. The characteristics, distribution, and migration of human populations on Earth's surface							✓		
10. The characteristics, distribution, and complexity of Earth's cultural mosaics							✓		
11. The patterns and networks of economic interdependence on Earth's surface						✓			
12. The processes, patterns, and functions of human settlement						✓			
13. How the forces of cooperation and conflict among people influence the division and control of Earth's surface									
14. How human actions modify the physical environment					✓				
15. How physical systems affect human systems						✓		✓	✓
16. The changes that occur in the meaning, use, distribution, and importance of resources									
17. How to apply geography to interpret the past	✓	✓				✓			
18. How to apply geography to interpret the present and plan for the future			✓	✓			✓		

Source: Geography Education Standards Projects. 1994. *Geography for Life: National Geography Standards 1994.* Washington, D.C.: National Geographic Research and Exploration.

9

Correlation to National Science Education Standards

Standard	Module 1		Module 2			Module 3		Module 4	
	Lesson 1	Lesson 2	Lesson 1	Lesson 2	Lesson 3	Lesson 1	Lesson 2	Lesson 1	Lesson 2
A. Science as inquiry			●♦	●♦	●♦	●♦	●♦	●♦	●♦
B. Physical science			●	●	●			♦	♦
C. Life science			●♦	●♦	●♦				
D. Earth and space science	●♦	●♦				●♦		●♦	●♦
E. Science and technology			●	●		♦			
F. Science in personal and social perspectives	♦	♦				●♦	●♦	♦	♦
G. History and nature of science	●♦	●♦	●	●	●♦	●♦	●♦	●♦	●♦

Key

● = grades K-4 standards

♦ = grades 5-8 standards

Standards reprinted with permission from National Science Education Standards. Copyright 1996 by the National Academy of Sciences. Courtesy of the National Academy Press, Washington, D.C.

Correlation to National Technology Standards

Standard	Module 1		Module 2			Module 3		Module 4	
	Lesson 1	Lesson 2	Lesson 1	Lesson 2	Lesson 3	Lesson 1	Lesson 2	Lesson 1	Lesson 2
1	✓	✓	✓	✓	✓	✓	✓	✓	✓
2			✓	✓	✓				
3								✓	✓
4	✓	✓	✓	✓	✓	✓		✓	✓
5									
6	✓	✓	✓	✓	✓	✓	✓	✓	✓

Source: National Education Technology Standards for Students. 2007. International Society for Technology in Education.

MODULE 1

Orientation to spatial thinking: World exploration

Introduction

Ferdinand Magellan's expedition circumnavigated the globe in the early sixteenth century. Along the way, Magellan and his crew stopped at many places, crossed several oceans, visited different continents, countries, and cities, and crossed major lines of latitude (parallels) and longitude (meridians) many times. This module puts students in the role of navigator, using GIS as a tool to get them from place to place all the way around the world. They will identify and find places along the route, learn about sailing in places with little wind (the doldrums) and lots of wind (trade winds), measure the distance between stops, zoom in and pan the map to get a better view of the route, change the flat map to a globe, and turn the globe as they travel.

Lesson 1: Magellan crosses the Atlantic Ocean

Lesson 2: Magellan crosses the Pacific Ocean

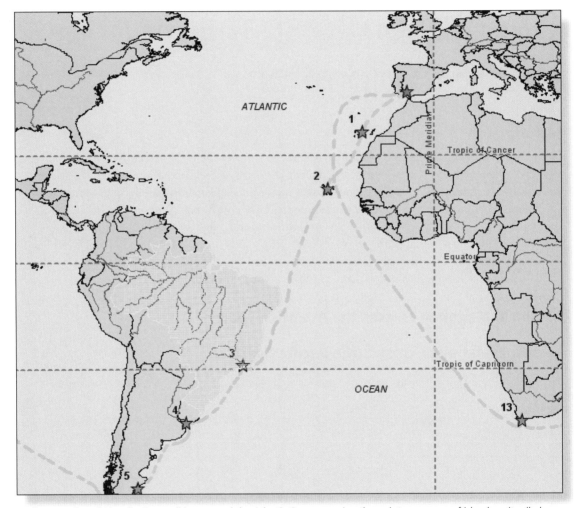

Explorer Ferdinand Magellan's expedition crossed the Atlantic Ocean, passing through two groups of islands as it sailed toward the east coast of South America.

Module 1: Lesson 1 ● ○

Magellan crosses the Atlantic Ocean

Lesson overview

Students will follow the path of Ferdinand Magellan on the first half of his expedition's circumnavigation of the globe—from Spain to the eastern coast of South America and across the Strait of Magellan to the Pacific Ocean. As they explore the world map, students will identify continents, countries, and cities along Magellan's route and work with major latitude and longitude lines. They will also learn about the doldrums and trade winds. They will interact with the map to change the scale and appearance of the map and answer relevant questions about Magellan's voyage.

While lesson 1 and lesson 2 can be done independently, students would benefit from doing both lessons sequentially.

Estimated time

Approximately 60 minutes

Materials

The student activity can be found in the student workbook or on the Software and Data CD. Install the teacher resources on your computer to access it.

Location: OurWorld_teacher\Module1\Lesson1
- Student activity: M1L1_student.pdf

Objectives

After completing this lesson, students will be able to do the following:

- Locate places on a world map
- Identify major lines of latitude and longitude
- Use cardinal directions to navigate from one place to another
- Recognize the difference between small-scale and large-scale maps

GIS tools and functions

🖿 Open a map file

🔍 Find a specific feature on the map

ⓘ Use the Identify tool to get information about a feature

🔍 Zoom to the geographic extent of a layer that is active (highlighted)

🔍 Zoom out to see more of the map (less detail)

📏 Measure distance on the map

- Turn layers on and off
- Activate a layer
- Label features
- Use the Pan Panel to pan (shift) the map (up, down, left, right)

National geography standards

Standard	K-4	5-8
1 How to use maps and other geographic representations, tools, and technologies to acquire, process, and report information	The characteristics and purposes of geographic representations—such as maps, globes, graphs, diagrams, aerial and other photographs, and satellite-produced images	The characteristics, functions, and applications of maps, globes, aerial and other photographs, satellite-produced images, and models
2 How to use mental maps to organize information about people, places, and environments	The location of Earth's continents and oceans in relation to each other and to principal parallels and meridians	How perception influences people's mental maps and attitudes about places
3 How to analyze the spatial organization of people, places, and environments on Earth's surface	The spatial concepts of location, distance, direction, scale, movement, and region	How to use spatial concepts to explain spatial structure
17 How to apply geography to interpret the past	How places and geographic contexts change over time	How people's differing perceptions of places, peoples, and resources have affected events and conditions of the past

M ● ○ ○ ○
L ● ○

Teaching the lesson

Introducing the lesson

Begin this lesson by reviewing these concepts:

- The major lines of latitude and longitude (e.g., equator, prime meridian, tropic of cancer, Arctic Circle) and how they are used to divide the world into hemispheres and zones
- Map scale and the difference between large-scale and small-scale maps
- How winds converge at the equator, creating the doldrums and making it hard for sailing ships to cross the equator
- The importance of Magellan's circumnavigation and what it proved about the world
- How the Spanish and the Portuguese divided the world between them in the Treaty of Tordesillas

Student activity

We recommend that you complete this lesson yourself before completing it with students. This will allow you to modify the activity to accommodate the specific needs of your students.

Teacher notes

- For younger grades, you can conduct the GIS activity as a teacher-led activity in which students follow along. You can lead students through the GIS steps and ask them the associated questions as a class.
- Ideally each student will have access to a computer, but students can complete the activities in groups or under the direction of a teacher.
- Throughout the GIS activity, students are presented with questions. The GIS activity sheets are designed so that students can mark their answers directly on these sheets. Alternatively, you can create a separate answer sheet.
- We recommend that students save their work as they progress through the GIS activity. Students can use either the Save command (to save their changes to the original map) or the Save As command (to save their changes to a new map). Please explain to students where and how they should save their work.

The following are things to look for while students are working on this lesson:

- As students work through the steps, are they focusing on the underlying geographic concepts (e.g., What are the spatial relationships between continents? How do major lines of latitude and longitude help us locate places on the earth?).
- Are students answering the questions in the GIS activity as they work through the steps?
- Are students able to use the legends to interpret the map layers?
- Are students aware of changes in scale as they zoom in and out on the map?

Concluding the lesson

- Engage students in a discussion about the observations and discoveries they made during their exploration of the world map.
- Ask students about their impressions of Magellan and his circumnavigation.
- Has this activity raised any questions that students would like to explore further?
- How can GIS help students to learn about world explorers and their discoveries?
- Has this activity changed students' ideas about maps?

Extending the lesson

- Have students research other circumnavigations and why Magellan's voyage was so important.
- Have students research the treaties between the Spanish and the Portuguese in the fifteenth and sixteenth centuries and how the line of demarcation changed over time.
- Have students research the navigation equipment that Magellan used during his voyage.

References

- http://www.ucalgary.ca/applied_history/tutor/eurvoya/
- http://www.nationalgeographic.com/volvooceanrace/geofiles/01/index.html
- http://www.expedition360.com/home/circumnavigation.htm

Student activity answer key

Answers appear in blue.

Module 1, Lesson 1

Magellan crosses the Atlantic Ocean

After Christopher Columbus found the New World in 1492, Spain and Portugal were eager to conquer and claim new lands. The two world powers decided to divide the world in half by drawing a line that ran through the Atlantic Ocean. Based on this line, Spain could claim new lands in the western half of the world, and Portugal could claim lands in the eastern half.

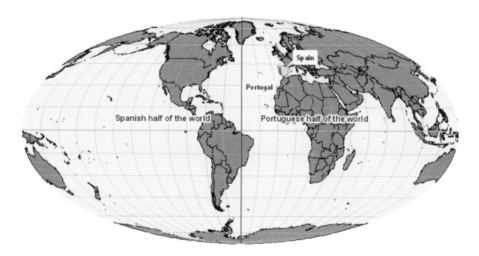

This shows the approximate location of the dividing line between the Spanish (western) half of the world and the Portuguese (eastern) half of the world.

Both countries wanted to claim the Spice Islands because of the cloves and other valuable spices that could be found there. But it was unclear where the Spice Islands were in relation to the dividing line because the islands were on the far side of the world from Europe.

Ferdinand Magellan, a Portuguese explorer born in 1480, led the first expedition to sail completely around the world. We call this a *circumnavigation*. Like Christopher Columbus, Magellan believed he could get to the Spice

Islands by sailing west. Magellan knew that he would have to sail around or through the Americas to do it.

Portugal turned down Magellan's plan, but he won support from the king of Spain. Magellan thought he could find a new route to the Spice Islands and perhaps claim those lands for Spain. In 1519, Magellan set sail with five ships and 270 men.

Let's follow Magellan's voyage and find out the new route he discovered.

Step 1: Start AEJEE.

1. Ask your teacher how to start the AEJEE software.

2. Click the **Maximize** button at the top right of the window. If you are not sure how to do this, ask your teacher.

    Now the AEJEE window fills your screen.

Step 2: Open the project.

1. Click the **Open** button. You see the Open window.

2. Choose **OurWorld1** and click **Open**.

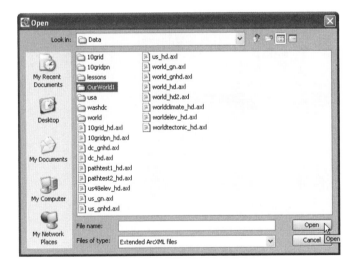

3. Choose **Module1** and click **Open**.

4. Choose **MagellanAtlantic.axl** and click **Open**.

A map of the world appears on your screen.

21

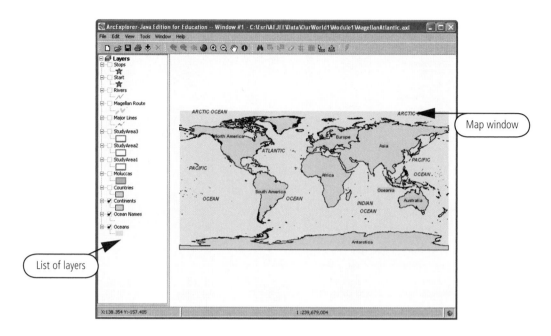

On the left side you see a list of layers. Layers are used to show geographic data on GIS maps. Each layer has a name and a legend. You can turn each layer on and off. At the top of the window, you see menus and buttons that you will use during this activity.

Q1 Look at the continents on the map and fill in the missing letters in the following statements.

a. Europe is touching the continent of A S I A.

b. The A T L A N T I C Ocean separates the continent of A F R I C A from South America.

c. The P A C I F I C Ocean separates the continent of North America from A S I A.

Step 3: Locate the Spice Islands, also known as the Moluccas.

1. Turn on the Moluccas layer by clicking the box next to Moluccas.

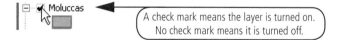

A check mark means the layer is turned on.
No check mark means it is turned off.

M ● ○ ○ ○
L ● ○

2. Look on your map for a small red area in the Far East.

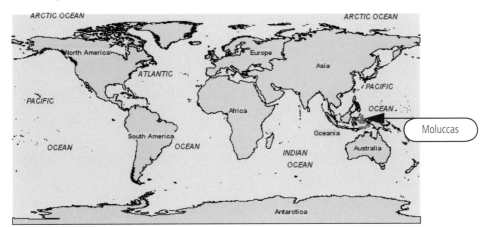

3. Click the **Continents** layer name to highlight it. It becomes active.

> **Note:** Making a layer active tells the GIS that this is the layer you want to work with or learn about.

4. Click the **Identify** tool. It allows you to get information about the active layer.

5. Click the **Moluccas** on the map.

The Identify Results window opens. On the left you see the name of the continent you clicked. On the right you see additional information about that continent.

Q2 What is the name of the continent where the Moluccas are found?

    a. Australia

    b. Oceania

    c. Asia

6. Close the **Identify Results** window.

23

7. Turn on the **Major Lines** layer by clicking the box next to its name. It shows the major lines of latitude and longitude (e.g., the equator and the prime meridian).

Q3 **What major line of latitude passes through the Moluccas? (Circle the correct answer.)**

   a. **Tropic of Cancer**

   b. **Equator**

   c. **Tropic of Capricorn**

   d. **Prime meridian**

8. Turn off the Major Lines layer by clicking the check mark next to its name.

Step 4: View Magellan's route.

You'll turn on different layers to see where Magellan sailed.

1. Turn on the **Countries** layer by clicking the box next to Countries in the list of layers. The outlines of all the countries appear on the map.

2. Turn off the **Continents** layer by clicking the check mark next to Continents.

3. Turn on the **Start** layer by clicking the box next to its name. A red star appears showing where Magellan started his voyage.

4. Turn on the **Magellan Route** layer by clicking the box next to its name.

The route Magellan followed around the world appears on the map.

M ● ○ ○ ○
L ● ○

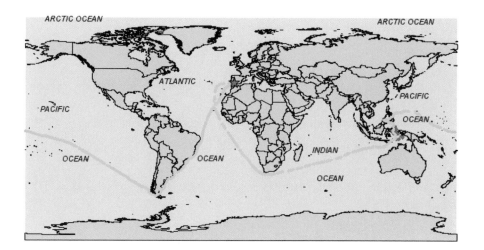

5. Click the **Start** layer name to highlight it. It becomes active.

6. Click the **Identify** tool.

7. Click the red star on the map. The Identify Results window opens.

(Q4) **What is the name of the city where Magellan began his voyage?** Seville

(Q5) **What country is this city in? (Hint: Look on the right side of the Identify Results window.)** Spain

8. Close the **Identify Results** window.

9. Turn on the **Major Lines** layer in the list of layers.

(Q6) **Magellan started at the red star and traveled west. How many times did he cross the equator to return to the starting point? (Circle the correct answer.)**

a. **One time**

b. **Two times**

c. **Three times**

d. Four times

25

A true circumnavigation requires crossing the equator at least once.

Q7 **Did Magellan spend more time north of the equator or south of the equator? (Circle the correct answer.)**

a. **North of the equator**

b. South of the equator

The equator divides the world in half between north and south. Each half is called a *hemisphere*.

Step 5:   Follow Magellan as he starts his voyage in September of 1519 and passes through two groups of islands.

You'll zoom in for a closer look at Magellan's route.

1.   Turn off the **Moluccas** layer by clicking the box next to its name.

2.   Turn on the **StudyArea1** layer.

3.   Click the **StudyArea1** layer name to highlight it. It becomes active.

4.   Click the **Zoom to Active Layer** button.

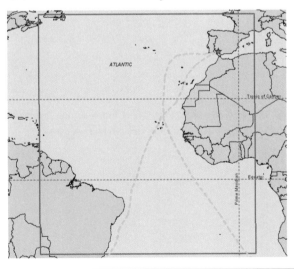

Q8 **How did your map change?**

    a. The map shows more detail (larger scale).

    b. The map shows less detail (smaller scale).

5. Turn off the **StudyArea1** layer.

6. Turn on the **Stops** layer.

Magellan passed through two groups of islands on his way to the equator.

Q9 **What direction did Magellan have to sail to reach these islands? (Circle the correct answer.)**

    a. **Northeast**

    b. **North**

    c. **Southeast**

    d. **Southwest**

7. Click the **Stops** layer name to highlight it. It becomes active.

 8. Click the **Identify** tool.

9. Click the first stop Magellan reached after leaving the start point.

 Q10 What is the name of the group of islands at this stop? Canary Islands

10. Close the **Identify Results** window.

11. Click the next stop.

Q11 What is the name of the group of islands at this stop? Cape Verde Islands

12. Close the **Identify Results** window.

27

Step 6: Magellan crossed the equator in October of 1519, one month after starting his voyage.

Near the end of October, Magellan's fleet neared the equator. The ships entered an area of calm water with no winds. This area is called the *doldrums*. The doldrums are located between 5 degrees north and 5 degrees south of the equator (see the orange band in the picture below). Magellan had to sail out of the doldrums in order to reach the *trade winds* south of the equator. Trade winds are warm, steady winds that blow all the time. They are located between 5 and 30 degrees north and south of the equator.

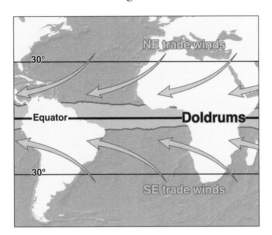

The *doldrums* are a zone of still air close to the equator. The *trade winds* are a zone of steady winds that blow from east to west.

In the doldrums, sailing ships often stood still, sometimes for days, making it frustrating for sailors. The weather there is hot and sticky. Only sudden thunderstorms with strong winds allowed ships to move.

 Based on what you know so far, why did it take Magellan three weeks to cross the equator and reach the trade winds? The doldrums are an area with no winds. Sailing in the doldrums would be very slow. Magellan would have to wait for a strong wind associated with a thunderstorm to move his ships.

 Look at the picture above. After crossing the equator, Magellan was trying to reach Brazil on the east coast of South America. Did the trade winds blow Magellan toward Brazil or away from Brazil?

a. Toward Brazil

b. Away from Brazil

Step 7: Magellan sailed along the coast of Brazil looking for a headland that would provide passage across the South American continent.

Magellan was told about a great headland near the southern tip of Brazil. (A headland is a place where the land is shaped like a "head" and sticks out into the ocean.) Magellan thought this headland could be a passage. He wanted to find it.

You'll find Brazil and select it on the map.

1. Click the **Find** button. The Find window opens.

2. In the Value box, type **Brazil** (make sure you type it correctly).

3. In the **Layers to Search** list, click **Countries**.

4. Click **Find**.

   The information appears on the right side of the window.

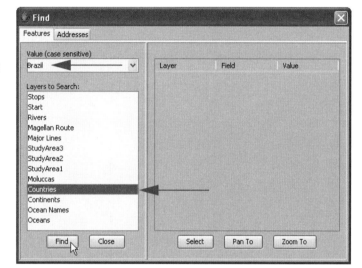

5. Click **Select** at the bottom of the window.

6. Close the **Find** window.

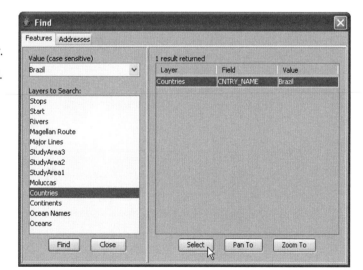

The country of Brazil is now highlighted in yellow on your map. You'll zoom out a little so you can see all of it.

7. Click the **Zoom Out** tool.

8. Place your mouse on the spot where Magellan's route to Brazil crosses the equator.

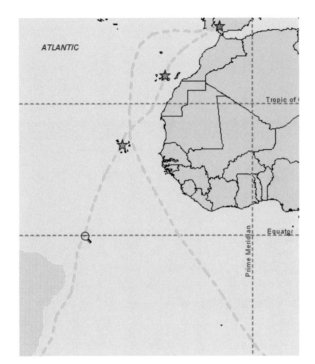

9. Click one time.

Q14 How did your map change? (Circle all the correct answers.)

 a. The map shows less detail (smaller scale).

 b. The map shows more detail (larger scale).

 c. The map shows a bigger area.

 d. The map shows a smaller area.

> **Note:** If you click in the wrong place, click the  Previous Extent button and try again.

10. Click the **Stops** layer name to highlight it.

11. Click the **Identify** tool.

30

(Q15) **What is the name of the place in Brazil where Magellan landed?** Rio de Janeiro

Magellan stopped here to make repairs and stock up on food. How far did Magellan travel to get here? You will use the Measure tool to find out.

12. Close the **Identify Results** window.

13. Click the **Measure** tool and choose **Nautical Miles** from the list.

Nautical Miles are slightly longer than regular miles.

- A regular mile is 5,280 feet in length
- A nautical mile is 6,076.1 feet in length

14. Click and *hold* your mouse on the starting place of Magellan's voyage.

15. Drag your mouse to Magellan's stop in Brazil. The gray box in the top left corner of your map shows the length of the line you are drawing.

16. Release the mouse button.

(Q16) **Look at the gray box. How many nautical miles did Magellan travel? (Hint: Use the number next to Total.)** 4, 1 8 8.7 4 7 nm (answers will vary)

Step 8: Magellan reached the headland on Christmas Day, 1519, three months after starting his voyage.

Magellan and his fleet left Brazil to sail along the east coast of South America, looking for the headland that he was told about. Just south of Brazil, they found it.

Before continuing, you will label the Stops.

31

1. Right-click **Stops** and choose **Properties**. This opens the Properties window.

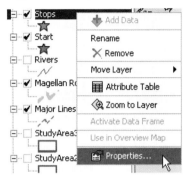

2. Click the **Labels** tab at the top.

3. Under **Label features using**, choose **StopNumber**.

4. Increase the **Size** to **12**.

5. Click the box next to **Bold** to check it.

6. Click the **Effects** button.

7. Click the box next to **Glow**.

8. Click **OK**.

9. Click **OK** again.

Now the stops are labeled with numbers that have a yellow glow around them.

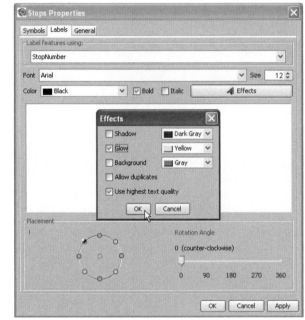

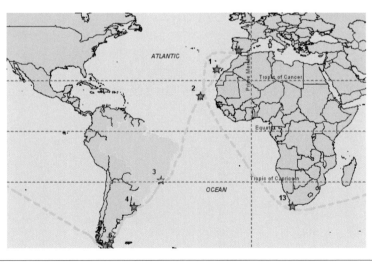

10. Look at the map and find Stop 4.

    Magellan sent one of his ships to explore this location. He hoped it was a place where ships could cross to the other side of the South American continent.

11. Turn on the **Rivers** layer by clicking the box next to its name.

12. Click the **Rivers** layer name to highlight it. It becomes active.

 13. Click the **Identify** tool.

14. Click the river that leads into the place where Magellan explored.

    (Q17) **What is the name of the river where Magellan's crew explored?** Parana

15. Close the **Identify Results** window.

    (Q18) **Why do you think Magellan couldn't sail his ships to the other side of the continent from this place? Possible answers include the following:**
    - **The river doesn't cross all the way to the other side of the continent.**
    - **It is hard to sail up a river in a large ship.**
    - **The channel might be too narrow.**
    - **The current might be too strong.**
    - **The wind might not be blowing the right direction.**
    - **The river might be too steep.**

    Magellan left the river in early February of 1520. His crew wanted to head north for the winter, but Magellan decided to sail south toward the tip of South America.

    To see the tip of South America, you will pan the map.

33

16. Click the **View** menu and choose **Pan Panel**.

You see a frame around your map. Each side of the frame has a small white arrow. By clicking one of these arrows, you can move or pan the map up (north), down (south), left (west), or right (east).

17. To pan south, click once on the white arrow at the bottom of the map.

Click once on the white arrow

Q19 **What happened to your map? Describe the changes in your own words.** The map moved up so that the area to the south is visible. The map is no longer centered on the equator. The southern tip of South America and the northern portion of Antarctica (below the Antarctic Circle) are now visible. The scale of the map did not change.

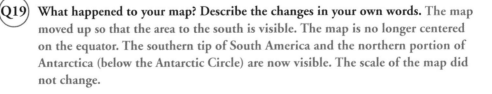

Step 9: Follow Magellan's route south along a deserted coast.

Magellan's ships faced strong winds and big waves as they sailed for about two months along the deserted coast. Bad weather seriously damaged three of Magellan's ships. By late March 1520, Magellan was forced to stop and spend the winter. He found a safe harbor at Stop 5.

(Q20) **What major line of latitude is south of the southern tip of South America? (Circle the correct answer.)**

　　a. **Equator**

　　b. **Tropic of Capricorn**

　　c. **Arctic Circle**

　　d. Antarctic Circle

(Q21) **As Magellan sailed south to Stop 5, do you think the temperature got warmer or colder?**

　　a. Colder

　　b. **Warmer**

1. Turn on **StudyArea2** by clicking the box next to its name.

2. Click the **StudyArea2** layer name to highlight it. It becomes active.

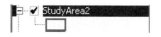

3. Click the **Zoom to Active Layer** button.

(Q22) **Look at your map. Do you see more detail or less detail?**

　　a. More detail

　　b. **Less detail**

4. Turn off **StudyArea2**.

5. Click the **Stops** layer name to highlight it.

6. Click the **Identify** tool.

7. Click Stop 5 (the star).

35

What is the name of Stop 5? San Julian

Magellan faced mutiny at this stop. A mutiny is when a group, like the crew on a ship, refuses to obey the person in charge, in this case, Magellan. Fortunately for Magellan, the mutiny failed and he punished the leaders.

8. Close the **Identify Results** window.

Step 10: Magellan discovered an ocean.

In October, 1520, more than a year after he began his voyage, Magellan and his crew found the headland they were looking for. They called it the Cape of the Eleven Thousand Virgins. This is Stop 6.

 Click Stop 6 to find out what this stop is called today. (Hint: Use the Identify tool.) Cape Virgenes

Two ships were blown west of Stop 6. Magellan thought they were lost. He found them in a deep, narrow channel of water (called a strait) that ran west-ward through the mountains. He followed this channel all the way across the tip of the South American continent.

 Click Stop 7 to find out what this famous passage is called today. (Hint: Use the Identify tool.) Strait of Magellan

M ● ○ ○ ○
L ● ○

1. Close the **Identify Results** window.

2. Turn on the **StudyArea3** layer. You can only see part of this layer now.

3. Click the **StudyArea3** layer name to highlight it. It becomes active.

4. Click the **Zoom to Active Layer** button.

   The scale of the map changed, so you can see a bigger area.

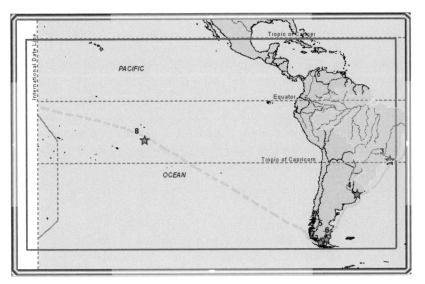

5. Turn off **StudyArea3**.

Q26 **What is the name of the ocean that Magellan reached on the west side of South America? (Circle the correct answer.)**

   a. **Atlantic Ocean**

   b. **Pacific Ocean**

Step 11: Save your work and exit AEJEE.

1. Ask your teacher how and where to save your work.

2. Click the **File** menu and choose **Exit**.

 Q27 **Write a short paragraph explaining what Magellan accomplished to this point. (Hint: What bodies of water and what major lines of latitude did he cross? What continents and countries did he visited?)** After leaving Spain, Magellan crossed the Atlantic Ocean sailing toward Brazil. When he got there, he searched for a place to cross the continent of South America. He eventually found this place and found an ocean on the other side—the Pacific Ocean. Along the way, he crossed the equator and the Tropic of Capricorn.

Conclusion

At this point, Magellan and his crew were the first Europeans to cross the South American continent and reach the ocean on the other side. They still needed to cross a big ocean to reach the continent of Asia and find the Spice Islands. With GIS, you have followed his path, gotten information about his stops, and measured distances along the way. In the next lesson, you will continue to follow Magellan's exploration.

M ● ○ ○ ○
L ● ○

An orthographic map, shaped like a globe, shows the route that Magellan's expedition followed as it sailed west across the Indian Ocean.

Magellan crosses the Pacific Ocean

Lesson overview

Students will follow the path of Magellan and his crew on the second half of their circumnavigation of the globe—from the west coast of South America, across the Pacific Ocean to the Spice Islands, and back to Spain. As they explore the world map, students will identify locations along Magellan's route using latitude and longitude values, measure distances between stops, and work with a map projection. Based on their exploration of the world map, students will answer questions about what Magellan's expedition accomplished.

While lesson 1 and lesson 2 can be done independently, students would benefit from doing both lessons sequentially.

Estimated time

Approximately 60 minutes

Materials

The student activity can be found in the student workbook or on the Software and Data CD. Install the teacher resources on your computer to access it.

Location: OurWorld_teacher\Module1\Lesson2
- Student activity: M1L2_student.pdf

Objectives

After completing this lesson, students will be able to do the following:

- Locate places on a world map
- Identify major lines of latitude and longitude
- Use latitude and longitude values to locate features
- Use cardinal directions to navigate from one place to another
- Recognize the difference between a small-scale map and a large-scale map

GIS tools and functions

Open a map file

Zoom to the geographic extent of a layer that is active (highlighted)

Add labels that appear when you place the mouse over a feature (called MapTips)

Zoom to the full geographic extent of all the layers in the map

Measure distance on the map

- Turn layers on and off
- Activate a layer
- Change the map projection
- Customize a map projection

National geography standard

Standard	K-4	5-8
1 How to use maps and other geographic representations, tools, and technologies to acquire, process, and report information	The characteristics and purposes of geographic representations—such as maps, globes, graphs, diagrams, aerial and other photographs, and satellite-produced images	The characteristics, functions, and applications of maps, globes, aerial and other photographs, satellite-produced images, and models
2 How to use mental maps to organize information about people, places, and environments	The location of Earth's continents and oceans in relation to each other and to principal parallels and meridians	How perception influences people's mental maps and attitudes about places
3 How to analyze the spatial organization of people, places, and environments on Earth's surface	The spatial concepts of location, distance, direction, scale, movement, and region	How to use spatial concepts to explain spatial structure
17 How to apply geography to interpret the past	How places and geographic contexts change over time	How people's differing perceptions of places, peoples, and resources have affected events and conditions of the past

M ● ○ ○ ○
L ● ● ●

Teaching the lesson

Introducing the lesson

Begin this lesson by reviewing these concepts:

- The major lines of latitude and longitude (e.g., equator, prime meridian, tropic of cancer, Arctic Circle) and how they are used to divide the world into hemispheres and zones
- How latitude and longitude are used to locate features precisely on the earth and on a map
- What a map projection is and how it changes the shape of the earth on a map
- The importance of Magellan's circumnavigation and what it proved about the world
- How the Spanish and the Portuguese divided the world between them

Student activity

We recommend you complete this lesson yourself before completing it with students. This will allow you to modify the activity to accommodate the specific needs of your students.

Teacher notes

- For younger grades, you can conduct the GIS activity as a teacher-led activity with students following along. You can lead students through the GIS steps and ask them the associated questions as a class.
- For older grades, ideally, each student will have access to a computer, but students can complete the activities in groups or under the direction of a teacher.
- Throughout the GIS activity, students are presented with questions. The GIS activity sheets are designed so that students can mark their answers directly on these sheets. Alternatively, you can create a separate answer sheet.
- We recommend that students save their work as they progress through the GIS activity. Students can use either the Save command (to save their changes to the original map) or the Save As command (to save their changes to a new map). Please explain to students where and how they should save their work.
- Software notes
  - Students should maximize the software window before opening an AXL file.
  - If students close AEJEE before completing the GIS activity, the MapTips function is not saved. Students will have to set MapTips again when they reopen the map.

The following are things to look for while students are working on this lesson:

- As students work through the steps, are they thinking about the underlying geographic concepts (e.g., How do major lines of latitude and longitude help them locate places on the earth? How does changing the map projection affect what they see on the map?)?
- Are students answering the questions in the GIS activity as they work through the steps?
- Are students able to use the legends to interpret the map layers?
- Are students aware of changes in the map as they change the map projection?

Concluding the lesson

- Engage students in a discussion about the observations and discoveries they made during their exploration of the world map.
- Ask students about their impressions of Magellan's expedition and its circumnavigation.
- Ask students to compare their experience working with a flat map of the world to their experience working with a map that looks like a globe.
- Has this activity raised any questions that students would like to explore further?
- How can GIS help students to learn about world explorers and their discoveries?
- Has this activity changed students' ideas about maps?

Extending the lesson

- Have students research other circumnavigations and why Magellan's voyage was so important.
- Students can research the treaties between the Spanish and the Portuguese in the fifteenth and sixteenth centuries and how the line of demarcation changed over time.
- Have students research the navigation equipment that Magellan used during his voyage.
- Ask students to pretend they are planning a circumnavigation. Using information from the GIS map, from books, and the Internet, have students write a description of the route they would take and the stops they would make along the way.
- In the MagellanPacific.axl file, have students turn on the Line of Demarcation layer and the Moluccas layer. Set the Projection to Orthographic and the Center Longitude to 150. Ask your students if the Moluccas are in the Spanish or the Portuguese half of the world.

References

- http://www.ucalgary.ca/applied_history/tutor/eurvoya/
- http://www.geocities.com/rolborr/magsuicide.html
- http://www.expedition360.com/home/circumnavigation.htm

Student activity answer key

Answers appear in blue.

Module 1, Lesson 2

Magellan crosses the Pacific Ocean

Ferdinand Magellan was the first European explorer to reach the Pacific Ocean when his expedition sailed through an opening, or strait, near the tip of South America in 1520. He named the ocean Mar Pacifico, which means peaceful sea. The strait, which connected the Atlantic and Pacific oceans, was later named for him.

At that point in his journey, Magellan and his fleet had been at sea for more than a year. He had lost two of his five ships. Now he would cross the Pacific Ocean with three ships, looking for the coast of Asia and the Spice Islands. However, he had no idea the Pacific Ocean would be so big!

Step 1: Start AEJEE.

1. Ask your teacher how to start the AEJEE software.

2. Click the **Maximize** button at the top of the window. Now the AEJEE window fills your screen.

Step 2: Open the project.

1. Click the **Open** button.

2. Navigate to your **OurWorld1\Module1** folder.

3. Choose **MagellanPacific.axl**.

4. Click **Open**.

    When the project opens, you see a map of the world showing Magellan's route on your screen.

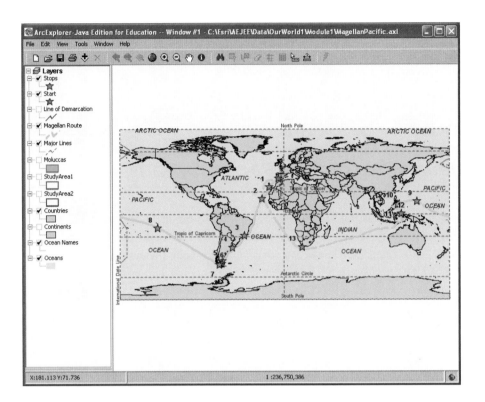

Notice the list of layers on the left side. Layers are used to show geographic data on GIS maps. Each layer has a name and a legend. You can turn each layer on and off. At the top of the window, you see menus and buttons that you will use during this activity.

Step 3: Follow Magellan on his Pacific voyage.

You are going to focus on the area where Magellan began to cross the Pacific Ocean.

1. Turn on the **StudyArea1** layer by clicking the box next to its name.

2. Click the **StudyArea1** layer name to highlight it. It becomes active.

3. Click the **Zoom to Active Layer** button.

The map redraws, and you are zoomed in to the Pacific Ocean.

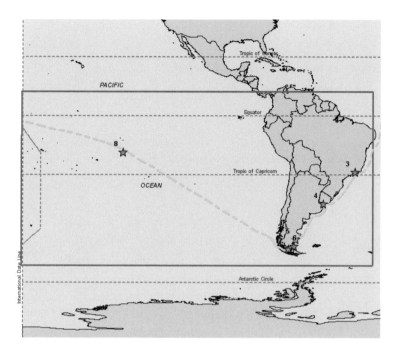

4. Turn off the **StudyArea1** layer by clicking the check mark next to its name.

Magellan hoped to find the coast of Asia, but the ships sailed for weeks without spotting land. Their food supplies rotted and began to disappear. After six weeks, his men began to die of scurvy, a disease caused by not having enough vitamin C. By mid-January 1521, more than a third of his men were too weak to walk, and they had only enough water for one sip a day.

Finally, on January 25, they spotted a small island at approximately 139 degrees west longitude and 14 degrees south latitude.

5. Place your mouse on the map while you look in the bottom left corner of the AEJEE window.

The numbers you see represent longitude (X) and latitude (Y). Longitude and latitude are used to identify the exact location of any place on the earth.

The X represents longitude—east or west of the prime meridian (a minus sign means west). The Y represents latitude—north or south of the equator (a minus sign means south).

6. Move your mouse until you find **X:-139 Y:-14**.

Module 1: Lesson 2

Q1 **What stop number is at this location?** Stop 8

Q2 **Is this location east or west of the prime meridian?**
    a. **East**
    b. West

Q3 **Is this location north or south of the equator?**
    a. **North**
    b. South

You can see the stop numbers, but you can't see their names. You will turn on MapTips so you can see the name of each stop when you place your mouse on it.

 7. Click the **MapTips** button. The MapTips window opens.

8. In the **Layers** list, scroll down and click **Stops**.

9. Under **Fields**, click **StopName**.

10. Click **Set MapTips**.

11. Click **OK**.

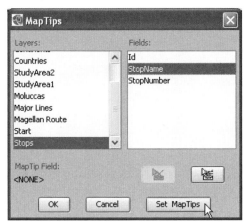

12. Hold your mouse over the red star at Stop 8 (don't click).

The name of the stop appears in a small box called a MapTip.

> **Note:** If you don't see the MapTip, move your mouse away from the red star, then move it back again.

Q4 **What is the name of the island at this stop?** P U K A P U K A

Step 4:   They left the island on January 28, 1521, and sailed onward.

To follow Magellan's route across the Pacific, you will change the shape of the map.

1. Click the **Tools** menu and choose **Projection** from the list.

2. In the **Select Coordinate System** window, click the **Custom** tab.

3. Click the drop-down arrow next to **Projection**.

4. Scroll down in the list until you see **Orthographic**, and click it.

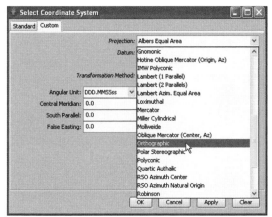

5. Click the **Center Longitude** box, and change the value from 0.0 to **-160.0**. (Be sure to include the minus sign!)

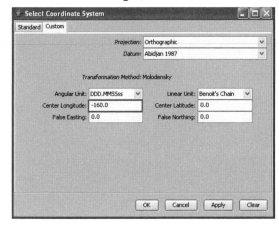

49

6.   Click **OK**.

 7.   Click the **Zoom to Full Extent** button.

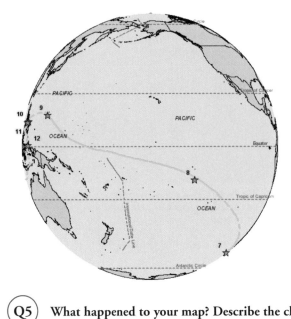

(Q5)   **What happened to your map? Describe the changes in your own words.** The map changed from flat (like a rectangle) to round (like a globe). The map is centered on the Pacific Ocean. The surface of the map appears curved.

Step 5:   Magellan's route missed most of the Pacific islands where he could have found food.

When Magellan left Stop 8, he decided to head west and slightly north. By March 4, about five weeks after leaving the island at Stop 8, Magellan's ship ran out of food, leaving only a dozen men strong enough to do any work.

On March 5, just when things seemed hopeless, they spotted land at Stop 9.

(Q6)   **What is the name of this stop? (Hint: Hold your mouse over the stop and read the MapTip.)** Guam

At this stop, native people carrying clubs, spears, and shields surrounded Magellan's ships. Magellan used crossbows, a type of bow and arrow, to kill some of them. These native people, called Chamorros, left in their canoes. Magellan stole their water and food and used cannons to bomb their village. Magellan's fleet sailed away as hundreds of Chamorros returned in their canoes to fight.

M ● ○ ○ ○
L ● ● ●

Before you turn the globe again, you will measure the distance Magellan traveled across the Pacific Ocean.

1. Click the **Measure** tool and choose **Nautical Miles**.

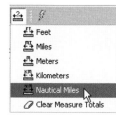

2. Click and *hold* down your mouse button on Stop 7.

3. Drag your mouse to Stop 10 or 11 (they are on top of each other).

4. Let go of your mouse button.

> **Note:** If you make a mistake, click the **Measure** tool 🗲 and choose **Clear Measure Totals**. Then try again.

You see three lines on the map. Each shows a different measurement. The three lines are explained in the box below.

> When you measure on a projected map in AEJEE, you see three different lines:
> * The red line is a straight line on the screen. It looks like the shortest distance between two points, but because the earth is round, it may not be the shortest distance in real life.
> * The blue line is a great circle route. It is the shortest distance between two points if you used a piece of string on a real globe to measure the distance.
> * The pink line is a rhumb line. This line follows the same compass direction for the entire trip.

Q7 **Look at the curved blue line on your map and the blue (geodesic) measurement in the gray box. What is the great circle distance that Magellan traveled on his voyage from South America to Asia? Write the answer here.**
8,0 6 4.6 7 7 nm (Answers will vary depending on placement of the mouse pointer)

51

5. Click the **Measure** tool and choose **Clear Measure Totals**.

Step 6: Magellan decided to sail to the Philippines instead of the Spice Islands.

At this point, Magellan decided to sail to the Philippines. To follow his route to the Philippines, you will turn the globe again.

1. Click the **Tools** menu and choose **Projection** from the list.

2. Click the **Custom** tab.

3. Click in the **Center Longitude** box, and change the number to **150.0** (**no** minus sign this time).

4. Click **OK**.

You will zoom in for a closer look.

5. Turn on the **StudyArea2** layer by clicking the box next to its name.

6. Click the **StudyArea2** layer name to highlight it. It becomes active.

7. Click the **Zoom to Active Layer** button.

8. Turn off the **StudyArea2** layer by clicking the check mark next to its name.

Q8 How did your map change? (Circle all the correct answers.)

a. The map shows more detail (larger scale).

b. **The map shows less detail (smaller scale).**

c. **The map shows a bigger area.**

d. The map shows a smaller area.

On March 16, 1521, four months after leaving the west coast of South America, Magellan reached the Philippine Islands.

Q9 What is the name of the island at Stop 10? (Hint: Use the MapTip.)
Homonhon, Philippines

Here the crew gathered and recovered from scurvy when they ate fruit with lots of vitamin C. On March 27, 1521, Magellan realized he was close to the coast of Asia when a large canoe with people brought him gifts made in China.

Step 7: Magellan further explored the Philippine Islands, from Stop 10 to Stop 11.

Q10 What is the name of the island at Stop 11? (Hint: Use the MapTip.)
Cebu, Philippines

In the days that followed, Magellan and his men spent their time trading, relaxing, and converting natives to Christianity. When some rajahs, or chiefs, refused to convert, Magellan decided to punish them but was instead killed during a battle with the native peoples.

His men left Magellan's body behind, and two ships set sail for the Spice Islands. They landed at Stop 12 in November of 1521, after a voyage of 820 days (more than 2 years) since they left Spain. They spent three months in the Spice Islands (also known as the Moluccas), stocking up on spices and preparing to return to Spain.

To see the route back to Spain, you will turn the globe again.

1. Click the **Tools** menu and choose **Projection** from the list.

2. Click the **Custom** tab.

3. Click in the **Center Longitude** box and change the number to **90.0** (**no** minus sign).

4. Click **OK**.

5. Click the **Zoom to Full Extent** button.

54

M ● ○ ○ ○
L ● ● ●

Q11 What happened to your map? Describe the changes in your own words.

The globe turned toward the east (to the right). The map is now centered on the Indian Ocean instead of the Pacific Ocean. The surface of the map still appears curved.

Q12 What ocean did they cross on their way back to Spain? (Circle the correct answer.)

a. Pacific Ocean

b. Indian Ocean

c. Atlantic Ocean

d. Arctic Ocean

6. Turn off **Countries** by clicking the check mark next to its name.

7. Turn on **Continents** by clicking the box next to its name.

You will find out how far it is from the Spice Islands to the southern tip of Africa.

8. Click the **Measure** tool and choose **Nautical Miles**.

9. Click and *hold* your mouse on Stop 12.

10. Drag your mouse to Stop 13.

11. Let go of your mouse button.

 Q13 Look at the blue geodesic measurement to find out the great circle distance from the Spice Islands to Africa.

Write it here: 6,3 0 3.2 6 1 nm (Answers will vary depending on placement of the mouse pointer)

 Q14 Compare your measurement in Q7 (across the Pacific Ocean) with the measurement you just made (across the Indian Ocean). Which ocean is the greatest distance across? (Circle the correct answer.)

a. Indian Ocean

b. Pacific Ocean

12. Click the **Measure** tool and choose **Clear Measure Totals**.

Step 8: Heading for home.

In February 1522, one of the two remaining ships began to sink. Half the men were sent onward under the command of Juan Sebastian del Cano. The others stayed behind in the Spice Islands to make repairs but were soon captured by pirates and hanged.

Del Cano led the crew into the Indian Ocean at the end of March 1522. Soon they were low on water and food and sick with scurvy again. Violent storms damaged the ship. Late in May of 1522, they managed to sail past the cape at the southern tip of Africa (Stop 13).

(Q15) **What is the name of this cape? (Hint: Use the MapTip.)** Cape of Good Hope

By July, they were completely out of food and water. Only 24 men were left, so they headed for the Cape Verde Islands.

To see the route, you will turn the globe one last time.

1. Click the **Tools** menu and choose **Projection** from the list.

2. Click the **Custom** tab.

3. Click in the **Center Longitude** box and change the number to **0.0** (**no** minus sign).

Select Coordinate System		

Standard | Custom

Projection: Orthographic

Datum: Abidjan 1987

Transformation Method: Molodensky

Angular Unit: DDD.MMSSss   Linear Unit: Benoit's Chain

Center Longitude: 0.0   Center Latitude: 0.0

False Easting: 0.0   False Northing: 0.0

OK   Cancel   Apply   Clear

4. Click **OK**.

5. Click the **Countries** layer name to highlight it. It becomes active even though it is not turned on.

6. Click the **Zoom to Active Layer** button.

7. Turn on the **Countries** layer by clicking the box next to its name.

8. Turn off the **Continents** layer.

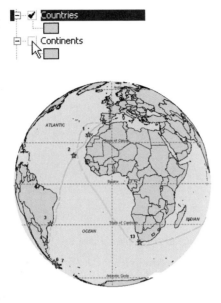

Do you remember where the Cape Verde Islands are? Magellan stopped there on his way to South America.

Q16 Which stop represents the Cape Verde Islands? (Hint: Use the MapTip.)

a. Stop 13

b. Stop 3

c. Stop 2

d. Stop 1

Q17 Which major lines of latitude did they cross on their way to this stop from the tip of Africa? (Circle all the correct answers.)

a. Antarctic Circle

b. Tropic of Capricorn

c. Equator

d. Tropic of Cancer

e. Arctic Circle

The Cape Verde Islands were in the half of the world claimed by the Portuguese, so del Cano and his Spanish crew were careful. When Portuguese troops arrived from a nearby fort, del Cano left in a hurry, leaving some of his men behind. He was afraid of losing the 26 tons of valuable spices (cloves and cinnamon) his ship was carrying.

In the first week of September 1522, almost three years after the voyage began, del Cano entered the bay that leads to the city of Seville, Spain—where the voyage started. Only 18 men had survived the journey. They were the first men known to travel all the way around the world. We call this a *circumnavigation*.

(Q18) **Which major line of latitude did they cross on their way from Stop 2 to their final destination? (Circle the correct answer.)**

    a. Antarctic Circle

    b. Tropic of Capricorn

    c. Equator

    d. Tropic of Cancer

    e. Arctic Circle

(Q19) **Write a short paragraph explaining what Magellan's expedition accomplished after leaving the South American continent. (Hint: What bodies of water and what major lines of latitude did the expedition cross? What continents and islands did the expedition visit?)**

After leaving the west coast of South America, the expedition accomplished the following:

- **Crossed the Pacific, Indian, and Atlantic oceans**
- **Crossed the Tropic of Capricorn and the equator four times**
- **Visited the continent of Asia**
- **Sailed around the southern tip of Africa (Cape of Good Hope)**
- **Visited the islands of Pukapuka, Guam, the Philippines, and the Spice Islands**

Step 9: Save your work and exit AEJEE.

1. Ask your teacher where and how to save your work.

2. Click the **File** menu and choose **Exit**.

59

Conclusion

Magellan's mission was to find a new route to the Spice Islands and claim them for Spain. Now you know that Magellan was killed and never reached the Spice Islands. Some of his crew made it there and returned with tons of spices, valuable enough to pay for the trip.

After Magellan's fleet visited the Spice Islands in 1521, Spain claimed that these islands were within their (western) half of the world. Eight years later, Portugal paid the Spanish 350,000 gold coins to give up their claim to the Spice Islands.

While few of Magellan's men survived, some of their most important accomplishments include the following:

- Being first to circumnavigate the globe
- Being first to navigate the strait in South America connecting the Atlantic and Pacific oceans
- Observing animals that were entirely new to Europeans (like a llama and a penguin)
- Discovering two galaxies (now called the Magellanic Clouds) in the night sky over the southern hemisphere
- Sailing 14,460 leagues (69,800 kilometers or 43,400 miles) around the earth, discovering its full extent (the actual extent or circumference of the earth is 40,075.16 kilometers or 24,901.55 miles)

MODULE 2

Classifying the world: The animal kingdom

Introduction

Animals are a big part of our lives. They fascinate us. Those who study animals classify them based on their physical characteristics, what part of the world they live in, what type of habitat they require, and what type of food they eat. Zoos are a place where people can get closer to see animals. This module puts students in the roles of zoo mapmaker, zoo tour planner, and zoo detective. Students will use GIS as a tool to create a zoo map, plan a zoo tour, and find the mystery animal that will be added to the zoo population. To create different maps, students will classify animals based on different characteristics to create different maps. They will query the GIS database to answer questions and make decisions. To identify the mystery animal, they will use GIS to search the world map.

Lesson 1: Mapping a zoo

Lesson 2: Touring a zoo

Lesson 3: Animals around the world

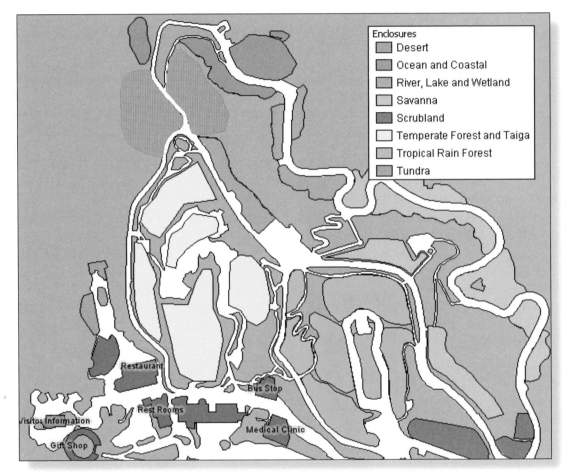

The colors on this map of the zoo show the type of habitat found in each of the enclosures.

Mapping a zoo

Lesson overview

Students will explore the concept of classification and how using categories can help make it easier to study certain things and ideas. Students will use GIS and work with a fictitious zoo to learn different methods of classification, create thematic maps, query data, and answer questions about what they see on the map.

While lesson 1, lesson 2, and lesson 3 can be done independently, students would benefit from doing all three lessons sequentially.

Estimated time

Approximately 60 minutes

Materials

The student activity can be found in the student workbook or on the Software and Data CD. Install the teacher resources on your computer to access them.

Location: OurWorld_teacher\Module2\Lesson1
• Student activity: M2L1_student.pdf

Objectives

After completing these lessons, students will be able to do the following:

• Classify animals using different characteristics and traits
• Interpret map symbology and create a map legend
• Work with attribute data
• Query data to find answers and make decisions

GIS tools and functions

⊕	Zoom in to a desired section of the map
↘	Set MapTips to display information about features without clicking
❶	Identify a feature on a map
▦	Use the Query Builder to select data based on certain criteria

Zoom to the full extent of all the layers

Add a layer to the map

Erase any selection made on the map features

- Turn layers on and off
- Activate a layer
- Change map symbology (legend)
- Query attribute data
- Label features
- Sort attributes
- Create a map layout for printing
- Select features

National geography standards

Standard	K-4	5-8
1 How to use maps and other geographic representations, tools, and technologies to acquire, process, and report information	How to display information on maps and other geographic representations	The characteristics, functions, and applications of maps, globes, aerial and other photographs, satellite-produced images, and models
2 How to use mental maps to organize information about people, places, and environments	The locations of places within the local community and in nearby communities	How to translate mental maps into appropriate graphics to display geographic information and answer geographic questions
18 How to apply geography to interpret the present and plan for the future	How people's perceptions affect their interpretation of the world	How varying points of view on geographic context influence plans for change

Teaching the lesson

Introducing the lesson

Begin the lesson by reviewing or discussing the following concepts:

- The classification of living things
- How classification can make it easier to study certain things such as animals, trees (deciduous or coniferous), books (fiction, nonfiction, biography, mystery, etc.), or food (dairy, meat, fruit, etc.)
- Map legends and symbols
- How features on maps have spatial and attribute information
- How maps can be used for:
  - Finding where things are

- Finding out how to get from one place to another
- Finding answers to questions

Student activity

We recommend that you go through this lesson yourself before completing it with students. This will allow you to become familiar with the content and the processes.

Teacher notes

- Explain the lesson to the students and ensure that they are aware of where to answer the questions asked.
- For younger grades, you can conduct the GIS activity as a teacher-led activity in which students follow along. You can lead students through the GIS steps and ask them the associated questions as a class.
- Students will each need a printed copy of the activity so they can answer questions throughout. They can mark their answers directly on the activity sheet. Alternatively, you can provide a separate answer sheet.
- Sometimes the names of the buildings will not all appear. This is due to the size of the map on your screen. Remember to maximize the AEJEE software before opening the project to help alleviate this problem.
- Ideally, each student will have access to a computer, but students can complete the lesson in small groups.
- Some questions do require classroom and/or group interaction. You can decide on the best way to handle these questions.
- Students will need access to a printer. Please explain how to use the print options.
- We recommend that students save their work as they progress through the GIS activity. Students can use either the Save command (to save their changes to the original map) or the Save As command (to save their changes to a new map). Please explain to students where and how they should save their work.

The following are things to look for while students work on this lesson:

- Are students thinking about the underlying geographic concepts as they work through the steps (e.g., how the animals are classified, how the zoo is organized, what attributes are available for the animals and habitats)?
- Are students answering the questions as they work through the steps?
- Are students experiencing any difficulties with the buttons, tools, mouse clicking, etc.?

Concluding the lesson

- Engage students in a discussion about animals around the world and where they live. Ask students to pick a specific country, continent, or region and research one or two of the animals that live there. They can present their findings to the classroom or in a report.
- Ask students to share any experiences or knowledge they have about maps. Bring in an atlas and show them different types of maps such as satellite images, street maps, and topographic maps.
- Engage students in a discussion about an endangered animal such as the panda, the gorilla, the elephant, or the tiger. Discuss how their habitats are being encroached on, and ask the students if they have any ideas on how to help endangered animals.
- Has this activity raised any questions that students would like to explore further?

- How can GIS help them learn about animals, organizing information, or finding answers to questions? GIS is used to track animal populations, provide information on habitat ranges, and analyze changes that occur in the environment.
- Have students do some Internet research on how zoos and wildlife preserves use GIS.

Extending the lesson

- Students can expand the zoo even further. Have them research natural habitats or regions and find animals that live in those habitats. They can add new enclosures to the zoo.
- Have students create a map of the classroom and classify the students into different categories such as students with pets, students with siblings, the months of their birthdays, their gender, and so on.
- Have students explore other living things and the ways they can be classified.
- Using the information from your map and from books or the Internet, write the script that a tour guide would give while escorting visitors through the zoo. The script could start like this:

    Welcome to our new zoo! Today we are going to start off by visiting the elephant enclosure. Elephants are endangered animals. Some elephants live in Africa, and other elephants live in Asia. Elephants use their trunks to drink water and pick up food (and so on).

References

- www.sandiegozoo.org
- www.torontozoo.com
- http://www.kidport.com/RefLib/Science/Animals/Animals.htm
- http://school.discovery.com/lessonplans/programs/animaladaptations/

Student activity answer key

Answers appear in blue.

Module 2, Lesson 1

Mapping a zoo

Animals are a big part of our life. Animals fascinate us, whether they live with us as pets or roam wild places on our planet. One exciting way to connect with animals from beyond our back yard is to visit the zoo! Zoological parks, or zoos, are a great way to bring people closer to animals. It is a chance for people to more deeply appreciate and understand how animals live and what they are like.

Zoos have been around for a long time. Queen Hatshepsut of Egypt had one about 3,500 years ago, and the Chinese emperor Wen Wang created a large zoo named the Garden of Intelligence about 3,000 years ago. Many leaders used zoos to show power and wealth. Zoos became popular starting about 500 years ago in the 1500s, when European explorers brought animals from the New World (the Americas) back to Europe.

Nowadays, zoos are a place where people can see animals from all over the world. Zoos are also a place where researchers can study animals more closely so that wildlife can survive where they normally live, in their natural habitat. This is important when evaluating how to help certain animals, especially endangered species, that otherwise might not survive on their own.

For this exercise, imagine that a new zoo is being built near your home. The animal enclosures are ready, and the animals are about to move into their new homes. With opening day quickly approaching, the zoo managers realized that they forgot to prepare a map for visitors! Your job will be to create a map that identifies the kinds of animals in the zoo, shows where they live, and provides information about the visitors' center, restrooms, and other facilities.

Step 1:   Start AEJEE.

1.   Ask your teacher how to start the AEJEE software.

2.   Click the **Maximize** button at the top right of the window. Now the AEJEE window fills your screen.

Step 2:   Open the project.

1.   Click the **Open** button.

2.   Navigate to the **OurWorld1\Module2** folder.

3.   Choose **Mappingazoo.axl**.

4.   Click **Open**.

A map of a zoo will be on your screen with other background information. On the left side of the map, you see a list of layers. Layers are used to show geographic data on a GIS map. Each layer has a name and a legend. You can turn each layer on and off. At the top of the map, you see buttons and tools that you will use during this activity.

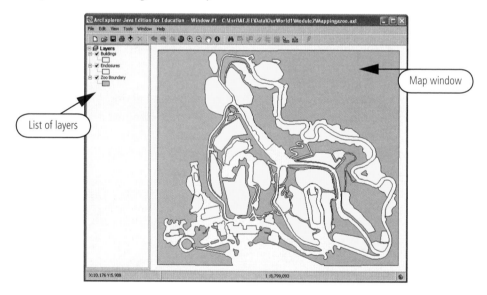

Step 3:   Finding things at the zoo.

First you should look at the information that is already on your map. Besides the animals themselves, there are many other features that make up a zoo.

Q1 Look on your map. Answer the following questions:

1. Can you find the restrooms?   Yes   No

2. Can you tell where the zebra lives?   Yes   No

3. Can you tell where the tropical rainforest is?   Yes   No

Q2 All your answers should be No. That's okay. Can you give two reasons why it is hard to find these things?

Answers (any reason that seems valid, including the three below):

1. Everything on the map is the same color.

2. There are no labels.

3. There is no legend.

Step 4: Finding the gift shop.

To help visitors find what they are looking for on the map, you should consider the kind of geographic data you can use to improve or enhance your map. Each layer has an attribute table full of information about the features in that layer.

Let's take a closer look at the attribute table for the Buildings layer.

1. Click the **Buildings** layer name to make it active. It will become highlighted.

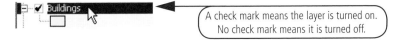

A check mark means the layer is turned on.
No check mark means it is turned off.

Note: Making a layer active tells the GIS that this is the layer you want to work with or learn about.

2. Right-click the **Buildings** layer and choose **Attribute Table**. The attribute table appears.

71

An attribute table is made up of fields (columns) and records (rows). This is where all the information about a layer is contained.

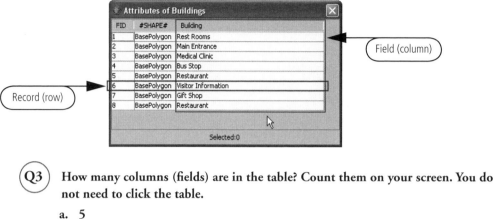

Q3 How many columns (fields) are in the table? Count them on your screen. You do not need to click the table.

a. 5

b. 3

c. 2

Q4 How many rows (records) are in the table? Count them on your screen. You do not need to click the table.

a. 8

b. 6

c. 9

Q5 Look under the Building field. Write down the name of two buildings that are listed in the attribute table.

Answer: Any of the following eight buildings are acceptable:

1. Main entrance

2. Restrooms

3. Medical clinic

4. Bus stop

5. Restaurant

6. Visitor information

7. Gift shop

3. Close the **Attributes of Buildings** window.

Now you can use the information you found in the attribute table to add more details to your map.

4. Right-click the **Buildings** layer and choose **Properties**.

5. On the **Symbols** tab, change the Color to **Gray**.

6. Click **Apply**.

7. Move the **Building Properties** window out of the way to see more of the map.

    Now the buildings are all colored gray. But you still don't know which building is which. You need to add some labels.

8. In the **Building Properties** window, click the **Labels** tab.

Module 2: Lesson 1

9. Under **Label features using**, choose **Building**.

10. Click **Effects**.

11. Put a check mark next to **Glow**.

12. Change the glow color to **Cyan**.

13. Click **OK**.

14. Click **OK** again.

The buildings are now labeled with their name. Visitors can find their way to the gift shop, the restrooms, and the restaurants.

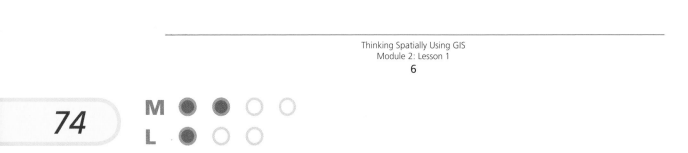

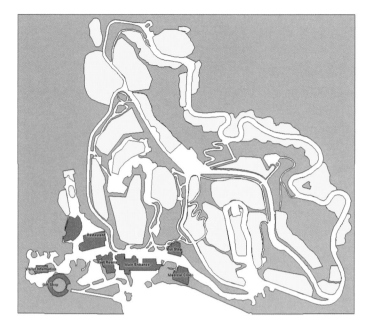

15. Ask your teacher about where and how to save your work. Save your project as **Mappingazoo1.axl**.

Step 5: Finding the tropical rainforest.

Zoos increasingly try to build their enclosures to imitate the natural habitat of animals that live in the wild.

We are going to make a map so that visitors can easily find the tropical rainforest that the zoo created and visit the animals living there. They will also be able to visit all the other types of habitats the zoo created along the way.

1. Click the **Enclosures** layer to make it active. It becomes highlighted.

2. Right-click the **Enclosures** layer and choose **Attribute Table**.

This attribute table contains all the information you have about the enclosures at the zoo. You can organize your table in different ways so that it is easier to read.

3. Right-click the field name **Habitat** and choose **Sort Ascending**. This will rearrange the habitats in alphabetical order.

Q6 **What kind of attribute information is available in the Habitat field?**
This field contains information on what type of habitat the animals live in.

Q7 **How many different types of habitat are there?**

a. 8

b. 6

c. 10

Q8 **List two habitats here:** Answers may include the following:

1. Savanna

2. Tropical Rain Forest

3. Desert

4. Scrubland

5. River, Lake, and Wetland

6. Temperate Forest and Taiga

7. Tundra

8. Ocean and Coastal

M ● ● ○ ○
L ● ○ ○ ○

4. Close the **Attributes of Enclosures** table.

 Q9 Look at your map. Can you see which enclosure represents which habitat on your map? Yes or No

5. Once again, your answer is No. If the habitats were different colors, they would be a lot easier to see. Let's set up symbols for the enclosures.

6. Right-click the **Enclosures** layer and choose **Properties**.

7. Under **Draw features using**, choose **Unique Symbols**.

8. Next to **Field for values**, choose **Habitat**.

This will assign a different color to each enclosure based on the habitat type it represents.

9. Next to **Color Scheme**, choose **Minerals**.

10. Keep the **Style** as **Solid Fill**.

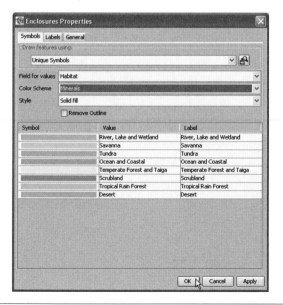

11. Click **OK**.

 Q10 **What happened to your map? Describe what happened in your own words.** Changing the symbols and colors made the map easier to read and understand. Now you can tell where each habitat is located. The legend shows the color that represents each habitat.

12. In the list of layers, look at the legend for the **Enclosures**.

 Q11 **What color is the tropical rainforest on your map?** This will vary from computer to computer.

13. In the legend for **Enclosures**, click the symbol that represents **Tundra**.

All the tundra will be highlighted in green.

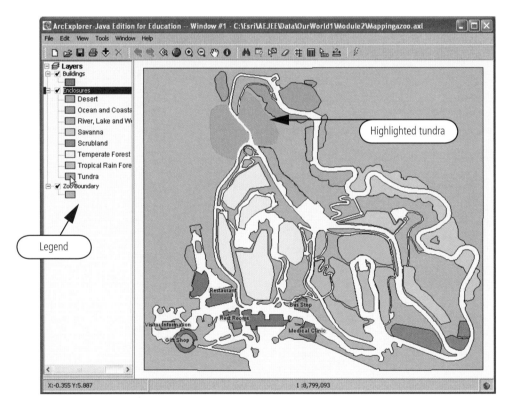

14. In the legend again, click the symbol for **Scrubland**.

    All the scrubland areas are now highlighted in green.

    Now you can see the relationship between the attribute table and the map. Using the attribute table, let's make your map more useful.

15. Click the **Clear All Selections** button.

16. Ask your teacher about where and how to save your work. Save your project as **Mappingazoo1.axl**.

Step 6: Adding the animals.

Sometimes, a zoo groups animals together based on their appearance. Other times, they are grouped based on their collective activities. Butterflies need to be in enclosed spaces (like a butterfly emporium) so they can't fly away. Monkeys need lots of trees and branches for swinging and climbing. Polar bears need a pool for swimming.

Your map is already full of great information, but you are still missing one key piece—the animals! Let's add them to the map.

1. Click the **Add Data** button.

2. Navigate to the **OurWorld1\Module2\Data** folder.

3. Choose **ZooAnimals.shp**.

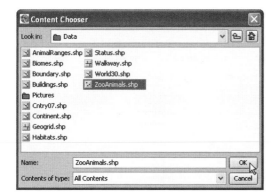

4. Click **OK**.

A new layer appears in the list on the left. A set of points appears on your map. The points represent different species of animals living in the zoo. Each animal is located in its proper habitat. Let's take a closer look at which animals live at the zoo.

5. Click the **ZooAnimals** layer name to make it active.

6. Click the **Identify** tool.

7. Click any animal point. This will bring up the **Identify Results** window.

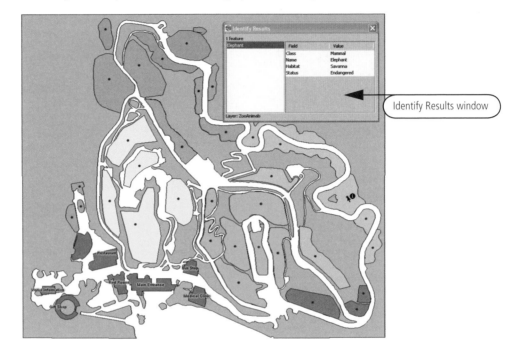

Identify Results window

(Q12) **Look at the data in the window and complete the following table:**
This will depend on what animal students choose. Here is an example:

Class	Mammal
Name	Zebra
Habitat	Savanna
Status	Endangered

8. Close the **Identify Results** window.

Using the Identify tool gives you the information for one animal at a time.

9. Right-click the **ZooAnimals** layer and choose **Attribute Table**.

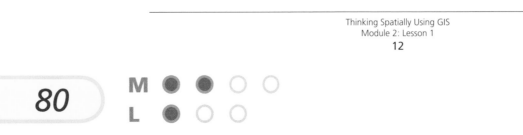

This is the attribute table associated with the animals. It gives you all the information about all the features at once.

10. Right-click the **Name** field and choose **Sort Ascending**. This will rearrange the animals in alphabetical order.

11. Scroll down the list of animals.

Q13 **Based on the attribute data in the table, list one way that you could classify (put into categories) the animals in the zoo.** Answers may include the following: by class, by habitat, by status, and by name.

Q14 **Based on the attribute data in the table, list one way that you could symbolize (represent with colors or symbols) the animals in the zoo.** Answers may include a picture of each animal and a different color for each animal.

Q15 **Can you think of another way that animals can be described or classified that is not in this table?** Answers may include the following:
- What type of food they eat
- What country/continent they live in
- What covers their body (fur, feathers, scales, etc.)
- Their size
- Their weight

12. Close the **Attribute Table**.

13. Right-click the **ZooAnimals** layer and choose **Properties**.

81

14. Under **Draw features using**, choose **Unique Symbols**.

15. Next to **Field for values**, choose **Habitat**.

16. Next to **Color Scheme**, choose **Minerals**.

17. Leave the **Style** as **Circle**.

18. Change the **Size** to **12**.

19. Click **OK**.

Now the animals are classified using habitat type. This is the same way the enclosures are classified. Does this give a visitor much extra information? Not really.

For example, look at the savanna habitat area. What information are you showing about the animals that live in that area? Only that they live in the savanna.

Let's try mapping the animals using different information. You can use the Class field so that visitors can see which animals are mammals, reptiles, birds, or insects.

20. Right-click the **ZooAnimals** layer and choose **Properties**.

21. Under **Draw features using**, choose **Unique Symbols**.

22. Next to **Field for values**, choose **Class**.

23. Leave the **Color Scheme** as **Random**.

24. Change the **Size** to **12**.

Now you will see four classes of animals with different colors assigned to them. You want to make sure that your colors stand out, so let's make some changes.

25. With the **Properties** window still open, click the color box next to **Bird**.

26. In the color table, choose a **dark blue** color.

27. Click **OK**.

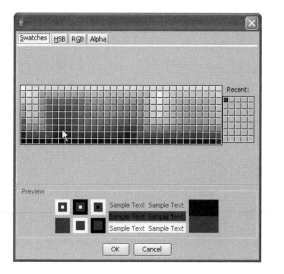

28. Click the color box next to **Insects and Spiders**.

29. In the color table choose a **red** color.

30. Click **OK**.

31. Click the color box next to **Mammal**.

32. In the color table, choose a **green** color.

33. Click **OK**.

34. Click the color box next to **Reptiles**.

35. In the color table, choose a **orange** color.

36. Click **OK**.

Finally, you need to label the animals so that visitors can see and find the location of specific animals in the zoo.

37. Click the **Labels** tab.

38. Under **Label features using**, choose **Name**.

39. Change the **Font** to **Verdana**.

40. Leave the size.

41. Put a check mark next to **Bold**.

42. In the **Placement** box, click the dot at the top of the circle.

43. Click **OK**.

Placement at the top of the circle

Now the animals are all labeled based on their names.

44. Ask your teacher about where and how to save your work. Save your project as **Mappingazoo1.axl**.

> **Note:** You may notice that some animals are missing a label. This can happen when two labels overlap. You can use the **Identify** tool to find the names of animals without labels.

Q16 Explain why your map is much more useful to visitors. What does your map show? What can you say about the colors and symbols? The map now shows the location of each animal in the zoo. It shows all the different types of animal habitats. The colors and symbols are clear and make it easy for visitors to find what they are looking for. Visitors can plan their day and make sure that they see all the animals they want. They can also find places like the gift shop, the restaurants, and the medical clinic.

Q17 Go back to step 3. Can you answer Yes to those questions now?

a. Yes

b. No

Step 7: You are ready for opening day!

Your new map is now informative, helpful, and also looks really nice. It's ready for visitors to use on opening day. All you need to do is print it.

1. Under the **View** menu, choose **Layout View**. Your map will appear centered on a "paper."

First you need to add a title to your map.

85

A 2. Click the **Add text** tool.

You see a white box that says **Right-click this text** in the middle of your map.

3. Click, hold, and drag the box to the top of your layout.

4. Right-click the box and choose **Properties**.

5. Click the white box and write the name of your zoo here (you can choose any name you want!).

6. Click **Change Properties**.

7. Change the **size** to **44**.

8. Click **OK**.

9. Click **OK** again.

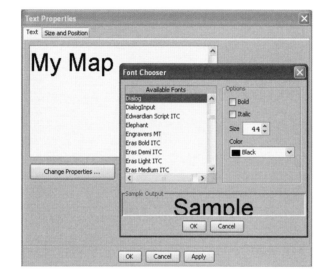

10. Click and hold on your title to move it around. Place it where you want it.

Now you need to add a legend to the map so that people can understand the way you classified things.

11. Click the map frame once. Blue dotted lines appear around it.

12. Click the **Add map legend** tool. Your legend appears on the map.

13. Click, hold, and drag your legend to the bottom right of your layout. You can resize the legend if you want.

14. Click the **Add image** button.

15. Navigate to **OurWorld1\Module2\Data\Pictures**.

M ● ● ○ ○
L ● ○ ○

16. Select ONE picture from the list of animals:
    - Elephant
    - Flamingo
    - Giraffe
    - Gorilla
    - Hippopotamus
    - Lion
    - Panda
    - Tiger
    - Zebra

17. Click **Open**.

    The animal will appear in a box on your map (it might appear right in the middle of your map).

18. Click, hold, and drag the box to the empty space on the top right corner of your page. You can resize the box if you want.

19. You can rearrange all the different parts, or elements, on your map until you are happy with the way it looks.

Step 8: Save your work, print your map, and exit AEJEE.

1. Ask your teacher about where and how to save your work. Save your project as **Mappingazoo1.axl**.

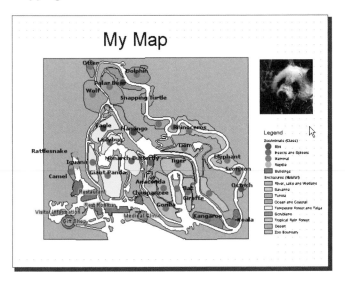

2. Ask your teacher how to print your map.

3. When you are finished printing your map, click the **File** menu and choose **Exit**.

Conclusion

Well done. You have now created a useful and informative map of your community's new zoo. Classifying features is not easy, and it is important to keep in mind what you are trying to show.

Visitors can use your map to plan a full day of fun at the zoo, including observing the animals, having lunch, and shopping. Looking at the map, visitors can learn a lot about the animals living in the zoo and understand how it is organized.

Zoos are constantly evolving and changing. Many animals are losing their natural habitats as the human population increases. Zoos and conservation groups around the world are using geographic information systems (GIS) to locate and monitor habitat loss. GIS is one of many tools that zoos use today to help people understand and appreciate the value of animals on our planet.

Students query the zoo map to find endangered mammals that live in savanna habitat.

Module 2: Lesson 2 ●●● ●●

Touring a zoo

Lesson overview

Students will create a custom tour of the animals in a zoo. They will examine the relationship between features and attribute data. They will perform simple and complex queries to answer questions about the information on the map and in the database. The will use the information they collect to make decisions about what animals to visit on the tour and draw the path that the tour will take.

While lesson 1, lesson 2, and lesson 3 can be done independently, students would benefit from doing all three lessons sequentially.

Estimated time

Approximately 60 minutes

Materials

The student activity and worksheets can be found in the student workbook or on the Software and Data CD. Install the teacher resources on your computer to access them.

Location: OurWorld_teacher\Module2\Lesson2
• Student activity and worksheets: M2L2_student.pdf

Objectives

After completing these lessons, students will be able to do the following:

• Work with tabular data
• Query data to find answers and make decisions
• Create a map legend for interpretation
• Explain different ways that animals can be classified

GIS tools and functions

Zoom in to a desired section of the map

Set MapTips to display information about features without clicking

Identify a feature on a map

Use the Query Builder to select data based on certain criteria

Zoom to the full extent of all the layers

Add a layer to the map

Erase any selection made on the map features

- Turn layers on and off
- Activate a layer
- Change map symbology (legend)
- Query attribute data
- Label features
- Sort attributes
- Create a map layout for printing
- Select features

National geography standards

Standard	K-4	5-8
1 How to use maps and other geographic representations, tools, and technologies to acquire, process, and report information	How to display information on maps and other geographic representations	The characteristics, functions, and applications of maps, globes, aerial and other photographs, satellite-produced images, and models
2 How to use mental maps to organize information about people, places, and environments	The locations of places within the local community and in nearby communities	How to translate mental maps into appropriate graphics to display geographic information and answer geographic questions
18 How to apply geography to interpret the present and plan for the future	How people's perceptions affect their interpretation of the world	How varying points of view on geographic context influence plans for change

Teaching the lesson

Introducing the lesson

Begin the lesson by reviewing or discussing the following concepts:

- How features on maps have spatial and attribute information
- The use of maps to answer questions and make decisions
- How guided tours are organized at zoos or other parks; examine ideas on which animals are popular, which animals are easy to see, and the kind of information that tours provide
- Visit your local zoo or wildlife preserve

The student activity

We recommend that you complete this lesson yourself before completing it with students. This will allow you to become familiar with the content and the processes.

Teacher notes

- Students must receive both handouts for this lesson—a table to be completed and a paper map of the zoo. Students should complete the first handout (the table) in its entirety and use the second handout (a map of the zoo) to draw their walking tour.
- For younger grades, you can conduct the GIS activity as a teacher-led activity in which students follow along. You can lead students through the GIS steps and ask them the associated questions as a class.
- Students will each need a printed copy of the activity so they can answer questions throughout. They can mark their answers directly on the activity sheet. Alternatively, you can provide a separate answer sheet.
- Sometimes the names of the buildings will not all appear. This is due to the size of the map on your screen. Remember to maximize the AEJEE software before opening your project to help alleviate this problem.
- Ideally, each student will have access to a computer, but students can complete the lesson in small groups.
- Some questions do require classroom and/or group interaction. You can decide on the best way to handle these questions.
- Students will need access to a printer. Please explain how to use the print options.
- We recommend that students save their work as they progress through the GIS activity. Students can use either the Save command (to save their changes to the original map) or the Save As command (to save their changes to a new map). Please explain to students where and how they should save their work.

The following are things to look for while students work on this lesson:

- Are students thinking about the underlying geographic concepts as they work through the steps (e.g., how the animals are classified, how the zoo is organized, what attributes are available for the animals and habitats)?
- Are students answering the questions as they work through the steps?
- Are students experiencing any difficulties with the buttons, tools, mouse clicking, etc.?

Concluding the lesson

- Ask students to share any experiences or knowledge they have about maps. Bring in an atlas and show them different types of maps such as satellite images, street maps, and topographic maps.
- Have students ever used a map at a zoo or amusement park? Ask students to compare their GIS map of a zoo to a paper map of a local zoo.
- Has this activity raised any questions that they would like to explore further?
- How can GIS be used to plan routes? Look at how GIS is used in evacuation planning, ambulance and fire truck routing, and even pizza delivery!

Extending the lesson

- Have students create a tour of the school. They should decide on what areas to visit and highlight and the most optimal way to visit them.
- Students can look at vacation tour maps and critique the tour—does it cover all the highlights of a city/town? Does it take into account some free time for tourists?
- Using the information from your map, books, and the Internet, write the script that a tour guide would use while escorting visitors through the zoo. The script could start like this:

> *Welcome to our new zoo! Today we are going to start off by visiting the elephant enclosure. Elephants are endangered animals. Some elephants live in Africa, and other elephants live in Asia. Elephants use their trunks to drink water and pick up food (and so on).*

References

- www.sandiegozoo.org
- www.torontozoo.com
- http://www.kidport.com/RefLib/Science/Animals/Animals.htm
- http://school.discovery.com/lessonplans/programs/animaladaptations/

Student activity answer key

Answers appear in blue.

Module 2, Lesson 2

Touring a zoo

The animal kingdom is quite large, with thousands of animal species identified around the world and more being discovered all the time. To make sense of all these species, scientists typically classify animals based on their physical characteristics. They start with a general classification and then get more detailed until they end up with a scientific name for the animal. For example, in the Linnaeus classification, the scientific name for a brown bear is *Ursus arctos*. This means that it has a backbone, is a mammal and a carnivore, and is part of the bear family.

Usually, it is easier to use common names to identify animals. In addition to their physical features, animals have many other characteristics: What country or area do they come from? What habitat do they live in? What kind of food do they like to eat?

For this lesson, the new zoo that has opened in your community needs your mapping expertise again. The zookeepers are so pleased with the map you made that they would like you to create a custom visitors tour. Sometimes, visiting a zoo can feel overwhelming because there is so much to see. In the last lesson, you used the animals' class (whether they were mammals, reptiles, or birds) to tell them apart. You will now explore some of the other information you have about the animals.

The zookeepers would like you to make sure visitors see a wide range of animals from different habitats and different classes. Here is a list of the characteristics of the animals that the zoo would like you to include in your tour:

- One reptile
- One bird
- One insect
- One animal that lives in the scrubland
- One mammal that lives in the tundra
- One mammal that lives in the savanna and is endangered

The tour also requires that you stop for lunch at the restaurant and make time to buy some souvenirs in the gift shop.

Let's get started on your tour map!

Step 1:   Start AEJEE.

1.   Ask your teacher how to start the AEJEE software.

2.   Click the **Maximize** button at the top right of the window. Now the AEJEE window fills your screen.

Step 2:   Open the project.

1.   Click the **Open** button.

2.   Navigate to the **OurWorld1/Module2/** folder.

3.   Choose **TouringtheZoo.axl**.

4.   Click **Open**.

A map of a zoo will be on your screen with other background information. It looks like the map you created in the previous activity. On the left side of the map, you see a list of layers. Layers are used to show geographic data on a GIS map. Each layer has a name and a legend. You can turn each layer on and off. At the top of the map, you see buttons and tools that you will use during this activity.

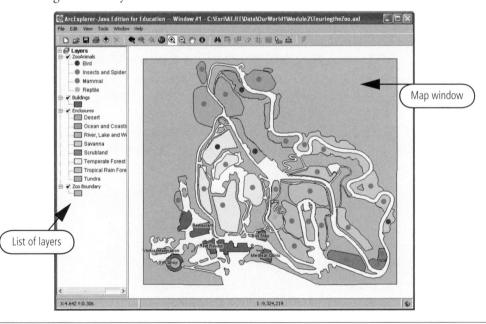

M
L

Step 3: Snakes and lizards.

Reptiles are cold-blooded animals with scaly skin. Their skin helps keep their bodies from drying out in the heat. They sit in the sun to get warm and then move to the shade to cool down. This class of animals lays eggs to have their babies. There are more than 6,500 species of reptiles in the world. That's a lot of reptiles!

Remember that your tour needs to include one reptile. Let's see what reptiles are at the zoo so we can choose which one to visit on our tour.

 Q1 **Look at the map legend. What is the color for reptiles?** Orange

A few reptiles live at the zoo. The problem is you don't know what kind of reptiles they are. That information is stored in the attribute table.

1. Click the **ZooAnimals** layer name to make it active.

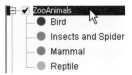

❶ 2. Click the **Identify** tool.

3. Click one of the **reptile symbols** in the zoo. This brings up the **Identify Results** window.

The **Identify Results** window contains information about the feature you selected. The information is coming from the attribute table.

 Q2 **Write down the name of the reptile in the Query answers column on your worksheet (worksheet 1).**

4. Close the **Identify Results** window.

5. Click another reptile symbol.

 Q3 Write down the name of this reptile in the Query answers column on your worksheet. Do this for all the reptiles. (Hint: There are 5 of them.)

6. Close the **Identify Results** window when you are done. You should have a list of all the reptiles in the zoo on your worksheet.

 Q4 Have you ever seen one of these reptiles in pictures or in real life? Write down the ones you have seen. Students should list which animals they have seen.

 Q5 Write down some things you know about one of the reptiles. Students should describe the animal they have seen. They can also include where they have seen it, what it looks like, if they touched it, and so on. There are many acceptable answers.

Step 4: Birds of a feather.

The zoo now wants visitors to see a bird. Birds are warm-blooded animals like mammals, but birds are covered in feathers. They have a backbone and a skeleton. Did you know that some of their bones are hollow to make it easier to fly? Birds lay eggs in nests, and when the babies hatch, the parents bring them food and care for them until they are old enough to fend for themselves. Of more than 8,800 known species of birds, the smallest is the hummingbird, and the largest is the ostrich.

You need to choose one bird for your tour. Do you know what kinds of birds are in the zoo? Let's find out.

 Q6 Look at the legend. What is the color for birds? Blue

Instead of using the Identify tool, you are going to select all the birds and then look at the attribute table to see what information you can find.

1. Look in the list of layers. Find the legend for ZooAnimals.

2. Click on the **Bird symbol** in the legend. (Hint: Click the circle.)

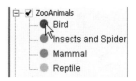

On the map, the bird symbols now have a green dot in the center that indicates they are selected.

3. Look at your map and locate the birds.

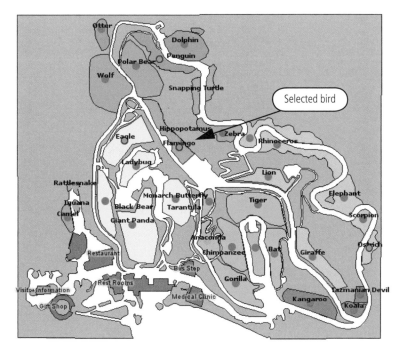

4. Right-click the **ZooAnimals** layer and choose **Attribute Table**. This will open the attribute table for the zoo animals. You will see that the selected features are highlighted in blue.

5. Right-click the **Name** field heading and choose **Sort Selected Data to Top**.

At the bottom of the window, the number of selected features is shown.

Q7 Write the names of all the birds under the Query answers column on your worksheet.

Q8 Scroll your table to the right. Look in the Status field. Which birds are listed as vulnerable (facing a high risk of extinction in the wild)?

1. Flamingo

2. Penguin

3. Eagle

Q9 Have you ever seen one of these birds in pictures or in real life? Write their names here. Students should list which animals they have seen.

Q10 Write down some things you know about one of the birds. Students should describe the animal they have seen. They can also include where they have seen it, what it looks like, if they touched it, and so on. There are many acceptable answers.

6.   Close the **Attribute Table**.

7.   Click on **Clear All Selections**.

Step 5:   Animals that live in the scrubland.

Scrublands have hot, dry summers and cool, moist winters. The few trees that grow there do not get very big. Although this seems like a harsh habitat, bobcats, leopards, owls, warthogs, and many other kinds of animals live in scrublands. In this zoo, several animals represent the scrubland. You need to choose one of them to add to your tour.

You have already used the Identify tool and selected features using the legend symbol. In a GIS, you can also build queries to get information from the attributes you have.

1.   Make sure the **ZooAnimals** layer is still active. (Hint: It should be highlighted in dark blue in the legend.)

2.   Click the **Query Builder** button. This opens the Query Builder window.

3.   Under **Select a field**, click **Habitat**.

4.   Click the = (equals) button.

5.   Under **Values**, click **Scrubland**.

Your query should look like this:
(**Habitat** = 'Scrubland')
You are asking the GIS to show you all the animals that live in the scrubland.

6.   Click **Execute**. A list of animals that live in scrubland appears in the lower half of the window.

Q11   **How many animals were selected?** Three

101

Q12 Write the names of all the animals under the Query answer column on your worksheet.

Q13 Have you ever seen one of these animals in pictures or in real life? Write their names here. Students should list which animals they have seen.

Q14 Write down some things you know about one of the animals. Students should describe the animal they have seen. They can also include where they have seen it, what it looks like, if they touched it, and so on. There are many acceptable answers.

7. Close the **Query Builder**.

You will see the same animals in the list highlighted in green on the map.

 8. Click **Clear All Selections**.

Step 6: Mammals and cold climates.

Mammals are animals that grow fur on their bodies and give birth to living young (they do not lay eggs). They also produce milk to feed their babies. Mammals are warm-blooded animals that regulate their body temperatures. However, that doesn't always mean they live in warm places.

Some mammals (and other animals) live in areas called the tundra. Tundra regions are found in the Arctic where temperatures drop as low at -76 degrees Fahrenheit! The summers can be quite warm, but they don't last long. Plants can still grow there, even though a layer of permafrost (ground that stays frozen all the time) covers the ground.

People are fascinated by animals that can survive in this kind of habitat. Let's find out what mammals at the zoo live in tundra regions.

1. Make sure the **ZooAnimals** layer is still active. (Hint: It should be highlighted in dark blue in the list of layers.)

2. Click the **Query Builder** button. This opens the Query Builder window. This time you will build a complex query where you ask the GIS more than one question at a time.

3. Click **Clear** to erase the old query.

4. Under **Select a Field**, click **Class**.

5. Click the = (equals) button.

6. Click **Mammal**.

7. Click **and**.

8. Click **Habitat**.

9. Click the = (equals) button.

10. Click **Tundra**.

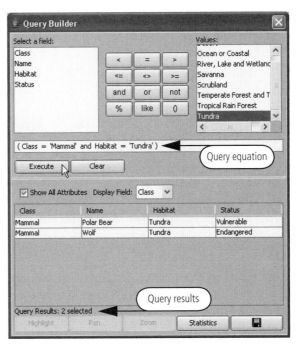

Your query should look like this:
(**Class = 'Mammal' and Habitat = 'Tundra'**)
You are asking the GIS to show you all the animals that are mammals and that live in the tundra.

11. Click **Execute**.

The list of animals that meet these criteria will appear in the list at the bottom of the window.

Q15 How many animals were selected? Two

Q16 Write the names of all the animals under the Query answers column on your worksheet.

Q17 Have you ever seen one of these mammals in pictures or in real life? Write their names here. Students should list which animals they have seen.

103

Q18 **Write down some things you know about one of the animals.** Students should describe the animal they have seen. They can also include where they have seen it, what it looks like, if they touched it, and so on. There are many acceptable answers.

12. Close the **Query Builder**. You will see the same animals in the list highlighted in green on the map.

13. Click the **Clear All Selections** button.

Step 7: Help save endangered species.

Thousands of animal species all over the world are considered endangered. There are several different levels of endangerment. Some animals have very low populations left and are considered endangered species because they are on the verge of disappearing from the earth (becoming extinct). Other animals are considered threatened species because they may become endangered in the future.

Zoos are often home to threatened and endangered species. Zoos are a way to help preserve their existence. Visitors and researchers appreciate being able to observe and study these animals. Many endangered animals are mammals that live in the African savanna. See if you can locate those animals in the zoo because you need to choose two of them to visit during the tour.

1. Make sure the **ZooAnimals** layer is still active. (Hint: It should be highlighted in dark blue in the table of contents.)

2. Click the **Query Builder** button. This opens the Query Builder window. You are once again going to build a complex query where you ask the GIS more than one question at a time.

3. Click **Clear** to erase the old query.

4. Under **Select a Field**, click **Class**.

5. Click the **=** (equals) button.

6. Click **Mammal**.

7. Click **and**.

8. Click **Habitat**.

Module 2: Lesson 2

9. Click the = (equals) button.

10. Click **Savanna**.

11. Click **and**.

12. Click **Status**.

13. Click =.

14. Click **Endangered**.

> Your query should look like this:
> (Class = 'Mammal' and Habitat = 'Savanna' and Status = 'Endangered').
> You are asking the GIS to show you all the animals that are mammals and that live in the savanna and that are endangered. That's a pretty complicated question!

15. Click **Execute**.

> The list of animals that are endangered mammals that live in the savanna will appear in the lower half of the window.

Q19 **How many animals were selected?** Two

Q20 **Write the names of all the animals under the Query answers column on your worksheet.**

Q21 **Have you ever seen one of these mammals in pictures or in real life? Write their names here.** Students should list which animals they have seen.

Q22 **Write down some things you know about one of the animals.**
Students should describe the animal they have seen. They can also include where they have seen it, what it looks like, if they touched it, and so on. There are many acceptable answers.

16. Close the **Query Builder**.

You will see the same animals in the list highlighted in green on the map.

17. Click the **Clear All Selections** button.

Step 8: Decide which animals to visit.

Now that you have made a list of animals that meet the requirements you got from the zoo, select the animals you want to visit and in which order. Don't forget to keep lunch and shopping in mind too!

1. Right-click the **ZooAnimals** layer and choose **Properties**.

2. Click the **Labels** tab.

3. Under **Label features using**, choose **Name**.

4. Change the **Font** to **Arial**.

5. Place a check mark next to **Bold**.

6. Change the **size** to **12**.

7. Click **OK**. Now all the animals are labeled by name.

8. Get your **worksheet 1** ready. Now you can decide which animals to include on your tour.

9. Look under the Query answers column. Choose one animal from each list and write it in the **Animal** column.

Example:

Class	Query answers	Animal (Write one animal from each query answer that you want to visit)	Order in tour
Reptile	1. Iguana		
	2. Anaconda		
	3. Snapping turtle	Snapping turtle	
	4. Rattlesnake		
	5. Crocodile		

10. Get out the map of the zoo that you created in the first activity (or use **worksheet 2**).

This map will look like the one on your computer screen. On the paper map you are going to draw the new tour route that visitors will take.

11. Look on the paper map and find the locations of all the animals you want to visit. These are the animals that you wrote in the **Animal** column on the worksheet.

12. Using the map, decide on the best order to visit the animals, stop for lunch, and shop in the gift shop. Make sure you go in a nice circle around the zoo. You don't want to cross back and forth.

13. Once you have decided the proper order, go to the **Order in tour** column on your **worksheet 1** and write the numbers 1 to 7 next to each stop in the order that you have chosen (you want to start at the Main Entrance).

14. Write the same numbers on your map in the correct locations.

15. On your paper map, draw a line that starts at the Main Entrance and joins all the points in order and ends at the Main Entrance.

Your map looks great! Now you can hand it out to visitors who request a custom tour of the zoo.

Step 9: Save your work and exit AEJEE.

1. Ask your teacher where and how to save your work.

2. Click the **File** menu and choose **Exit**.

Q23 From the tour you created, can you tell what additional animals will be seen along the way (besides the ones you selected)? List three of them here and two attributes (characteristics) about each animal. The answers will vary from student to student depending on the tour route they choose to create.

Animal	Attribute 1	Attribute 2

Conclusion

A great deal of planning and research goes into creating a custom tour of a zoo or any other tourist attraction. You want to make sure that the visitors get the most information and best look at the zoo.

All the animals have different characteristics and come from unique places. Identifying the animals and their characteristics on a map can make it more interesting to study them and learn more about their lifestyles.

M ● ● ○ ○
L ● ● ● ○

Module 2, Lesson 2

Touring a zoo

Worksheet 1

This handout will help you organize and keep track of your answers in the lesson. Enter the answers to your queries in the spaces below. In some cases, there will be more animals than you need. Choose one animal from each query answer to use in your tour.

Class	Query answers	Animal (Write one animal from each query answer that you want to visit)	Order in tour
Reptile	1. Iguana	The answers in this column will vary from student to student.	The answer in this column will vary from student to student.
	2. Anaconda		
	3. Snapping turtle		
	4. Rattlesnake		
	5. Crocodile		
Bird	1. Flamingo		
	2. Penguin		
	3. Eagle		
	4. Ostrich		
Animals that live in the Scrubland	1. Kangaroo		
	2. Koala		
	3. Tasmanian devil		
Mammals that live in the Tundra	1. Polar bear		
	2. Wolf		
Mammals that live in the Savanna and are endangered	1. Zebra		
	2. Elephant		
You need stop for lunch at the restaurant			
You need to stop at the gift shop			

Worksheet 2: Touring the Zoo

Name: _____

Date: _____

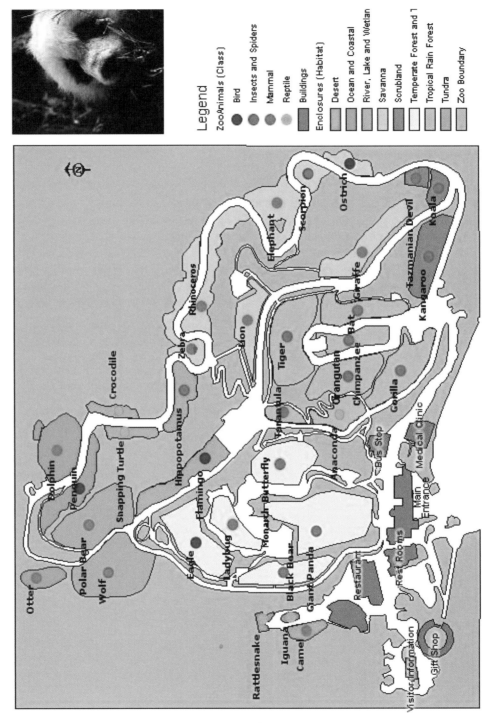

110

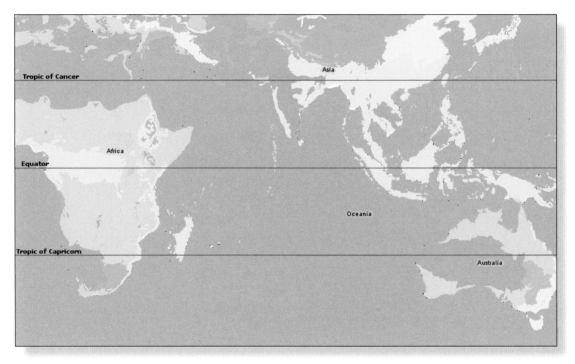

The colors on this map show the biomes or natural regions in the southeastern portion of the world where students search on the map for a mystery animal.

Module 2: Lesson 3 ●●●

Animals around the world

Lesson overview

Students will explore the concept of classification and animal habitats at a global level. Using data and information they have acquired, students will locate and identify a new animal to be introduced to the zoo. They will work with data from the World Wildlife Fund (WWF), including ecoregions, biomes, continents, and countries. Students will learn different ways that the world can be divided—geographically with lines of latitude and longitude or physically using biomes and ecoregions.

While lesson 1, lesson 2, and lesson 3 can be done independently, students would benefit from doing all three lessons sequentially.

Estimated time

Approximately 60 minutes

Materials

The student activity and worksheets can be found in the student workbooks or on the Software and Data CD. Install the teacher resources on your computer to access them.

Location: OurWorld_teacher\Module2\Lesson3
• Student activity and worksheets: M2L3_student.pdf

Objectives

After completing this lesson, students will be able to do the following:

• Use classification and querying skills to derive information from the map
• Describe different ways to classify the world—biomes and ecoregions
• Interpret map symbology
• Work with tabular data
• Create a map legend for interpretation

GIS tools and functions

⊕ Zoom in to a desired section of the map

⬚ Set MapTips to display information about features without clicking

ⓘ Identify a feature on a map

⬚ Use the Query Builder to select data based on certain criteria

⬤ Zoom to the full extent of all the layers

- Turn layers on and off
- Activate a layer
- Change map symbology (legend)
- Query attribute data
- Label features
- Sort attributes
- Create a map layout for printing
- Select features

National geography standards

Standard	K-4	5-8
5 That people create regions to interpret Earth's complexity	The concept of a region as an area of Earth's surface with unifying geographic characteristics	The elements and types of regions
8 The characteristics and spatial distribution of ecosystems on Earth's surface	The components of ecosystems	The local and global patterns of ecosystems
14 How human actions modify the physical environment	That the physical environment can both accommodate and be endangered by human activities	The consequences of human modification of the physical environment

Teaching the lesson

Introducing the lesson

Begin this lesson by reviewing these concepts:

- The classification of living things and how organizing and categorizing animals and plants can make them easier to study and analyze
- The geographic and functional organization of the world; the division of the world into continents, countries, states, and into natural regions such as climate zones, vegetation zones, and so on
- Map legends and symbols
- How features on maps have both spatial and attribute information

M ⬤ ⬤ ◯ ◯
L ⬤ ⬤ ⬤

- Endangered areas of the world such as the trees of the tropical rainforests or Antarctica's ice caps
- Endemic species of animals—they are unique to their own place or region and are not found naturally anywhere else

The student activity

We recommend that you go through this lesson yourself before completing it with students. This will allow you to become familiar with the content and the processes.

Teacher notes

- Explain the lesson to the students and ensure that they are aware of where to answer the questions asked.
- For younger grades, you can conduct the GIS activity as a teacher-led activity in which students follow along. You can lead students through the GIS steps and ask them the associated questions as a class.
- Students will each need a printed copy of the activity so they can answer questions throughout. They can mark their answers directly on the activity sheet. Alternatively, you can provide a separate answer sheet.
- Students need to complete the worksheet to unravel the mystery.
- Ideally, each student will have access to a computer, but students can complete the lesson in small groups.
- Some questions do require classroom and/or group interaction. You can decide on the best way to handle these questions.
- We recommend that students save their work as they progress through the GIS activity. Students can use either the Save command (to save their changes to the original map) or the Save As command (to save their changes to a new map). Please explain to students where and how they should save their work.

The following are things to look for while students work on this lesson:

- Are students thinking about the underlying geographic concepts as they work through the steps?
- Are students answering the questions as they work through the steps?
- Are students experiencing any difficulties with the buttons, tools, mouse clicking, etc.?

Concluding the lesson

- Engage students in a discussion about some of the animals that they are mapping in the activity. Look at the ranges of the animals. Where do they live? Why? What are the characteristics of their habitat and how does that make it ideal for a particular animal to live there?
- Ask students to share experiences of seeing animals in another country or another place.
- Discuss how classification made the lesson either more difficult or easier for them to complete.
- Has this activity raised any questions that they would like to explore further?

Extending the lesson

- Have students create a layout for this map and identify where the kiwi comes from.
- Students can research the kiwi in more detail and learn things such as how big their eggs are and why kiwi don't have wings.
- Students can choose another animal on the list and look at its range of habitat. Research that animal and explain why it lives where it does.
- Create a table of "official U.S. state animals." Research each state's animal and create an attribute table that lists the animal and some of its characteristics.
- Research the WWF and other organizations with similar missions: to help preserve the world's ecosystem.

References

- www.sandiegozoo.org
- www.wwf.org

Student activity answer key

Answers appear in blue.

Module 2, Lesson 3

Animals around the world

The zoo in your community is so popular and successful that it has decided to expand. After careful research, zookeepers have decided to add an exotic animal to the zoo population. They are holding a contest for visitors to guess what the new animal will be. You will use skills you have learned in classification and analysis to find what part of the world the new animal is from and then identify it.

To help you get started, the zoo has provided a list of possible animals. A list of clues will help you choose the correct answers. You will combine information you have in multiple layers of maps to find your answer.

Step 1: Start AEJEE.

1. Ask your teacher how to start the AEJEE software.

2. Click the **Maximize** button at the top right of the window. Now the AEJEE window fills your screen.

Step 2: Open the project.

1. Click the **Open** button.

2. Navigate to **OurWorld1\Module2**.

3. Choose **MysteryAnimal.axl**.

4. Click **Open**.

    A map of the world appears on your screen with some other background information. A legend to the left contains a list of the layers you will be working with. A series of buttons and tools will help you perform GIS functions.

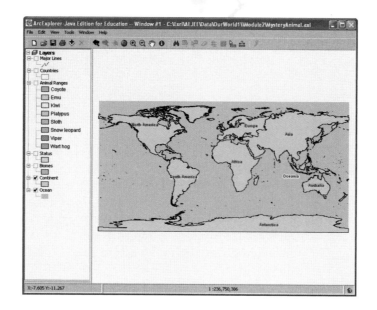

Step 3: Clue #1—The mystery animal lives in the Eastern Hemisphere.

The Eastern Hemisphere is the area of the world found east of the prime meridian and west of the International Date Line.

1. Turn on the **Animal Ranges** layer by placing a check mark next to the layer name.

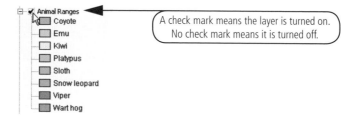

A check mark means the layer is turned on. No check mark means it is turned off.

The legend lists the names of new animals the zoo is considering. The map shows where those animals live in the world. A different color represents each animal. The same list of animals appears in your "**Animal list**" worksheet.

2. Turn off the **Animal Ranges** layer by clicking the check mark next to the layer name.

The animal that you are looking for lives east of the prime meridian (0° longitude).

3. Turn on the **Major Lines** layer by clicking the box next to the layer name.

This is a layer showing the major lines of latitude in the world.

4. Locate the intersection of the prime meridian and the North Pole.

5. Click the **Zoom In** tool.

6. Click and hold your mouse where the prime meridian and the North Pole meet.

7. Drag a box around the entire Eastern Hemisphere and stop at the far southeast tip.

8. Release your mouse button.

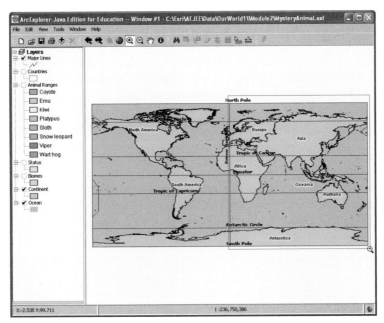

You will now have a closer look at the continents in the Eastern Hemisphere.

> **Note:** If you make a mistake, click on the **Zoom to Full Extent** button 🌐 and start over.

9. Turn on the **Animal Ranges** layer.

10. Click on the **Animal Ranges** layer name to highlight it. It will become active.

11. Click the **Identify** tool.

12. Click each animal range located in the Eastern Hemisphere. The name of the animal will appear in the Identify Results window.

Q1 On your "Animal list" worksheet, place an X under the animals located in the Eastern Hemisphere (either part of or all of their range is in the hemisphere).
Viper, Warthog, Emu, Kiwi, Platypus, Snow Leopard

13. Turn off the **Animal Ranges** layer.

Step 4: Clue #2—The mystery animal lives near the Tropic of Capricorn.

The **Tropic of Capricorn** is a major line of latitude located at approximately 23.5°S. This is the most southerly latitude where the sun can appear directly overhead at noon. This happens during the December solstice—about December 21—the official change from the spring season to the summer season (remember, the seasons are reversed in the Southern Hemisphere!).

Q2 Look at your map. Is the Tropic of Capricorn located in the Northern Hemisphere (north of the equator) or in the Southern Hemisphere (south of the equator)?

   a. **Northern Hemisphere**

   b. **Southern Hemisphere**

Q3 Look at your map. What continents does the Tropic of Capricorn cross? Circle all that apply.

   a. **Africa**

   b. **Europe**

   c. **North America**

   d. **Australia**

1. Click the **Zoom In** tool.

2. Click and hold your mouse at the intersection of the equator and the prime meridian.

3. Drag a box to the far southeast corner of your map.

120

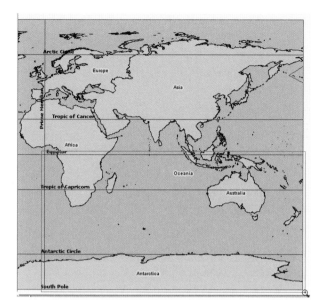

4. Release your mouse button.

You are now zoomed in to the Southern Hemisphere, east of the prime meridian.

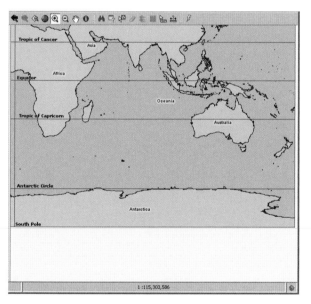

5. Turn on the **Animal Ranges** layer.

6. Make sure the **Animal Ranges** layer is still active. (Hint: It should be highlighted in dark blue in the legend.)

7. Click the **Identify** tool.

8. Click each animal range that is visible in the Southern Hemisphere.

 Q4 On your "Animal list" worksheet, place an X next to the animals that are located in the Southern Hemisphere (either part of or all of their range must be in the hemisphere). Warthog, Emu, Kiwi, Platypus

9. Turn off the **Animal Ranges** layer.

Step 5: Clue #3—The mystery animal lives in temperate mixed broadleaf forest areas.

A biome is an area that has similar animals, plants, and climate. A biome is a way of classifying the world into natural regions so that it's easier for people to study and research them.

Let's use the information you have to make a map showing the world's biomes.

 1. Click the **Zoom to Full Extent** button. This will zoom the map out to the whole world so that you can see all the biomes.

2. Turn off the **Continent** layer.

3. Turn on the **Biomes** layer.

4. Right-click the **Biomes** layer and choose **Properties**.

5. Under **Draw features using**, choose **Unique Symbols**.

6. Under **Field for values**, choose **BIOME_DESC**. (This field contains the descriptive name for each biome.)

7. Change the **Color scheme** to **Pastels**.

8. Click the box next to **Remove Outline**.

9. Click **OK**.

   Now each biome is identified with a different color.

Biomes Properties

Symbols | Labels | General

Draw features using:

Unique Symbols

Field for values	BIOME_DESC
Color Scheme	Pastels
Style	Solid fill

☑ Remove Outline

Symbol	Value	Label
	Tropical & Subtropical Grasslan...	Tropical & Subtropical Grasslan...
	Tropical & Subtropical Conifero...	Tropical & Subtropical Conifero...
	Rock and Ice	Rock and Ice
	Temperate Grasslands, Savan...	Temperate Grasslands, Savan...
	Temperate Conifer Forests	Temperate Conifer Forests
	Deserts & Xeric Shrublands	Deserts & Xeric Shrublands
	Montane Grasslands & Shrubla...	Montane Grasslands & Shrubla...
	Tropical & Subtropical Dry Broa...	Tropical & Subtropical Dry Broa...
	Lake	Lake
	Boreal Forests/Taiga	Boreal Forests/Taiga
	Flooded Grasslands & Savannas	Flooded Grasslands & Savannas
	Mangroves	Mangroves
	Tundra	Tundra
	Mediterranean Forests, Woodl...	Mediterranean Forests, Woodl...
	Tropical & Subtropical Moist Br...	Tropical & Subtropical Moist Br...
	Temperate Broadleaf & Mixed ...	Temperate Broadleaf & Mixed ...

OK | Cancel | Apply

Module 2: Lesson 3

10. Right-click the **Biomes** layer and choose **Attribute Table**.

11. Locate the **BIOME_DESC** field heading.

12. Look down the list until you see **Temperate Broadleaf & Mixed Forests**.

13. Click the record (row) containing **Temperate Broadleaf & Mixed Forests**.

   It will become highlighted in blue in your table.

123

Attributes of Biomes

FID	#SHAPE#	BIOME_DESC
1	BasePolygon	Boreal Forests/Taiga
2	BasePolygon	Deserts & Xeric Shrublands
3	BasePolygon	Flooded Grasslands & Savannas
4	BasePolygon	Lake
5	BasePolygon	Mangroves
6	BasePolygon	Mediterranean Forests, Woodlands & Scrub
7	BasePolygon	Montane Grasslands & Shrublands
8	BasePolygon	Rock and Ice
9	BasePolygon	Temperate Broadleaf & Mixed Forests
10	BasePolygon	Temperate Conifer Forests
11	BasePolygon	Temperate Grasslands, Savannas & Shrublands
12	BasePolygon	Tropical & Subtropical Coniferous Forests
13	BasePolygon	Tropical & Subtropical Dry Broadleaf Forests
14	BasePolygon	Tropical & Subtropical Grasslands, Savannas & S...
15	BasePolygon	Tropical & Subtropical Moist Broadleaf Forests

Selected:1

14. Close the **Attribute Table**.

You will see all the temperate broadleaf and mixed forest regions outlined in bright yellow on your map. This shows the relationship between the attribute data and the map.

Temperate broadleaf and mixed forests are typically located between the tropics and the poles. The temperature changes in these areas are not extreme, but there are distinct warm and cool seasons. This area is home to a variety of animal species, including brown bears, owls, leopards, and ladybugs.

Now that you understand a bit about biomes, return to the southeastern part of the world and continue your search for the mystery zoo animal.

15. Click the **Zoom In** tool.

16. Find the intersection of the equator and the prime meridian.

17. Click, hold, and drag a box that starts there and covers the southeastern part of the world.

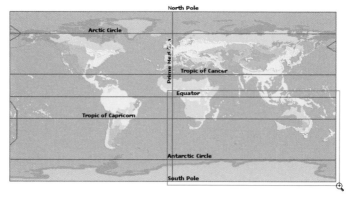

Module 2: Lesson 3

124

M
L

You are now zoomed back to the Southern Hemisphere.

Remember, you are looking for temperate broadleaf and mixed forests in the southeastern portion of the world near the Tropic of Capricorn.

18. Turn on the **Countries** layer. The countries will appear with a black outline over the biomes.

19. Click the **MapTips** button.

20. Under **Layers**, choose **Countries**.

21. Under **Fields**, choose **CNTRY_NAME**.

22. Click **Set MapTips**.

23. Click **OK**.

24. Place your pointer over the countries (don't click!), and the names of countries will appear.

Q5 What two countries in this area have temperate broadleaf and mixed forest? (Hint: You are looking for the yellow highlighted areas south of the equator and east of the prime meridian.)
1. Australia
2. New Zealand

 25. Click the **Zoom In** tool.

26. Click, hold, and drag a box around these two countries.

> **Note:** If you make a mistake, click on the **Zoom to Previous Extent** button ↩ and start over.

27. Click the **Biomes** layer to make it active. It becomes highlighted.

 28. Click the **Clear All Selections** button.

You have narrowed down the countries to Australia and New Zealand.

29. Turn off the **Biomes** layer.

30. Turn on the **Animal Ranges** layer.

31. Right-click the **Animal Ranges** layer and choose **Properties**.

32. Click the **Labels** tab.

33. Under **Label features using**, choose **COMM_NAME**.

34. Click **OK**.

Q6 On your "Animal list" worksheet, place an X under the names of the animals found in these two countries. Emu, Kiwi, Platypus

Step 6: Clue #4—The mystery animal is widespread across its country but lives mostly in endangered or critical areas.

Now you will look at the conservation status of the ecoregions in each country.

Ecoregions are smaller than biomes. They each have their own plant, animal, and climate characteristics. The World Wildlife Fund has defined 825 ecoregions on earth. One type of information that the WWF records about ecoregions is their conservation status. There are three levels of classification:

- Stable or intact—Species and ecosystems are abundant and not at any significant risk.
- Vulnerable—Where the area is at *high* risk for the extinction of species and reduction of the ecosystems.
- Critical or endangered—Where species in this area are at a *very high* or *extremely high* risk of becoming extinct.

1. Turn off the **Animal Ranges** layer.

2. Turn on the **Status** layer by placing a check mark next to the layer name.

3. Right-click the **Status** layer and choose **Properties**.

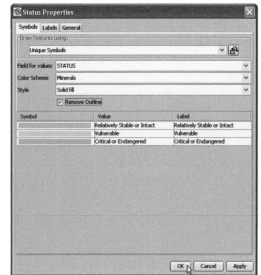

4. Under **Draw features using**, choose **Unique Symbols**.

5. Under **Field for values**, choose **STATUS**.

6. Change the **Color Scheme** to **Minerals**.

7. Place a check mark next to **Remove Outline**.

8. Click **OK**.

Q7 Look at your legend. Which areas of Australia and New Zealand are critical or endangered?

   a. The coastal areas.

   b. The central areas.

127

Q8 Which country is mostly covered by critical or endangered areas?

    a. Australia

    b. New Zealand

9. Turn on the **Animal Ranges** layer.

10. Click the **Animal Ranges** layer name to make it active.

 11. Click the **Identify** tool.

12. Click the range for the animal that lives in the country you just identified.

Q9 What is the name of the mystery animal? Kiwi

Step 7: Placing the mystery animal.

Now that you have identified the mystery animal, you need to decide where it should live in the zoo.

Use the map of the zoo provided by your teacher (worksheet 2). Think about the following when you are choosing a new home for this animal:

1. Should you put the animal near other animals with similar characteristics or in a separate area?
2. What kind of habitat do you think it would like? Remember the biome that it was living in (go back to step 5.)
3. Will it be seen on the tour that you created?

Q10 Draw the new enclosure on your map—be sure to label the enclosure with the name of the zoo's newest resident. The location of the enclosure and size will vary from student to student.

Q11 Write a brief paragraph explaining the location you chose for the animal. Look for solid explanations perhaps based on information students have learned in the past or in this lesson.

M
L

Step 8: Save your work and exit AEJEE.

1. Ask your teacher where and how to save your work.

2. Click the **File** menu and choose **Exit**.

Conclusion

The kiwi is a small flightless bird that lives only in New Zealand. New Zealand has more flightless birds than any other country. Some people believe the birds there didn't need to fly because they didn't have many enemies to fly away from.

The kiwi bird is about the size of a chicken. Kiwis usually mate for life and typically lay only one egg at a time. The reason for this is that a kiwi egg can be up to 10 times larger than a chicken egg! Kiwis have an excellent sense of smell and are the only bird with nostrils at the end of its beak!

Unfortunately, the kiwi population in New Zealand faces threats from an increase in predators, human influence, and loss of habitat. Numerous organizations aim to help the kiwi begin to thrive again in New Zealand.

Module 2, Lesson 3

Animals around the world

Worksheet 1: Animal list

Class	Viper	Wart hog	Sloth	Emu
Lives in the Eastern Hemisphere?	x	x		x
Lives in the Southern Hemisphere?		x		x
Lives in the two countries found in step 5?				x

Class	Coyote	Kiwi	Platypus	Snow Leopard
Lives in the Eastern Hemisphere?		x	x	x
Lives in the Southern Hemisphere?		x	x	
Lives in the two countries found in step 5?		x	x	

M
L

Worksheet 2: Touring the Zoo

Name: _____   Date: _____

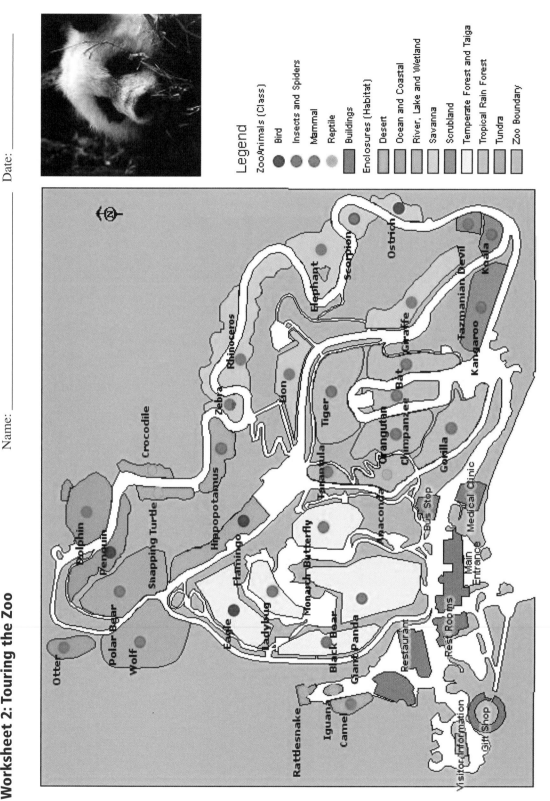

Legend

ZooAnimals (Class)
- Bird
- Insects and Spiders
- Mammal
- Reptile

Enclosures (Habitat)
- Buildings
- Desert
- Ocean and Coastal
- River, Lake and Wetland
- Savanna
- Scrubland
- Temperate Forest and Taiga
- Tropical Rain Forest
- Tundra
- Zoo Boundary

Module 2: Lesson 3

131

MODULE 3

Recognizing patterns: People and places

Introduction

When Europeans settled the land now known as the United States, they started in the east and moved west. Their settlements formed different patterns on the land. The population of the United States has grown quickly during the past several hundred years. Keeping track of the nation's population and documenting its characteristics have become the job of the U.S. Census Bureau. This module puts students in the role of U.S. population analyst. Using GIS, they will uncover population patterns of the past and the present. Students will identify early settlement patterns and compare them to other kinds of information included in separate GIS layers. They will compare graduated color maps showing where people live, where people move, and some of their characteristics. Students will also compare their own states to the rest of the country.

Lesson 1: Early settlement patterns of the United States

Lesson 2: Patterns of a growing population

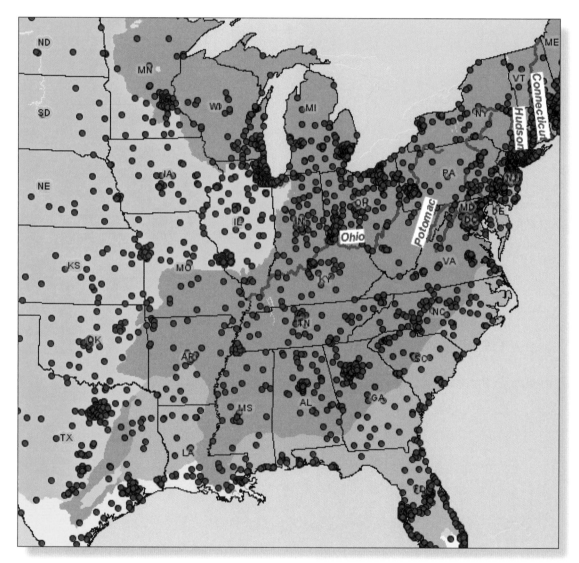

Students compare the pattern of cities that developed in the eastern United States with the locations of rivers and the types of vegetation.

Module 3: Lesson 1

Early settlement patterns of the United States

Lesson overview

Students will explore the distribution of human settlements in the United States. They will start on the Atlantic Coast and follow the movement of people westward to the Pacific Coast. As they follow the westward migration, students will use GIS to examine patterns in the distribution of settlements and will use different data layers to compare the locations of settlements to rivers, landforms, elevation, rainfall, and vegetation. Students will also examine some of the incentives for people to move west and some of the barriers they encountered.

Estimated time

Approximately 60 minutes

Materials

The student activity can be found in the student workbook or on the Software and Data CD. Install the teacher resources on your computer to access them.

Location: OurWorld_teacher\Module3\Lesson1
* Student activity: M3L1_student.pdf

Objectives

After completing this lesson, students will be able to do the following:

* Recognize and describe spatial patterns in the distribution of human settlements across the United States
* Identify areas of the United States that have a common spatial pattern
* Give possible reasons to explain settlement patterns
* Compare settlement patterns to other patterns on the land, such as landforms and land cover (vegetation)

GIS tools and functions

🗁 Open a map file

⊕ Zoom to the geographic extent of a layer that is active (highlighted)

🖳 Add labels that appear when you place the mouse over a feature (called MapTips)

✛ Add a layer of data to the map

⊖ Zoom out to see more of the map (less detail)

◄ Zoom to the previous extent

🖳 Use the Query Builder to select data

✋ Use the Pan tool to shift (move) the map in any direction

Functions

- Turn layers on and off
- Activate a layer
- Label features
- Change symbols
- Use the Pan Panel to pan (shift) the map (up, down, left, right)

National geography standard

Standard	K-4	5-8
11 The patterns and networks of economic interdependence on Earth's surface	The factors that influence the location and spatial distribution of economic activities	How changes in technology, transportation, and communication affect the location of economic activities
12 The process, patterns, and functions of human settlement	The factors that affect where people settle	Human events that led to the development of cities
15 How physical systems affect human systems	The ways in which the physical environment provides opportunities for people	How the characteristics of different physical environments provide opportunities for or place constraints on human activities
17 How to apply geography to interpret the past	That geographic contexts influence people and events over time	How people's differing perceptions of places, peoples, and resources have affected events and conditions in the past

M ● ● ● ○
L ● ● ○

Teaching the lesson

Introducing the lesson

Begin this lesson by reviewing these concepts:

* The distribution of features over space forms spatial patterns
* The three types of spatial patterns: regular (evenly spaced), clustered (clumped close together), and random (irregularly space or scattered)
* The progression of European settlement in America from east to west
* How natural features (landforms, rivers) influenced settlement patterns
* The importance of economic opportunities in advancing settlement and westward migration
* The first European settlements in America, such as Jamestown, Plymouth, and Boston
* The role of waterways such as the Chesapeake Bay, Hudson River, and Ohio River in settlement and trade
* The Appalachian Mountains as a barrier to settlement
* The Mississippi River system

Student activity

We recommend that you complete this lesson yourself before completing it with students. This will allow you to modify the activity to accommodate the specific needs of your students.

Teacher notes

* For younger grades, you can conduct the GIS activity as a teacher-led activity in which students follow along. You can lead students through the GIS steps and ask them the associated questions as a class.
* Ideally each student will have access to a computer, but students can complete the activities in groups or under the direction of a teacher.
* Throughout the GIS activity, students are presented with questions. The GIS activity sheets are designed so that students can mark their answers directly on these sheets. Alternatively, you can create a separate answer sheet.
* We recommend that students save their work as they progress through the GIS activity. Students can use either the Save command (to save their changes to the original map) or the Save As command (to save their changes to a new map). Please explain to students where and how they should save their work.

The following are things to look for while students are working on this lesson:

* As students work through the steps, are they thinking about the underlying geographic concepts (e.g., Are features distributed evenly or unevenly over space? Do the features form patterns on the map? What factors might explain the patterns?)
* Are students answering the questions in the GIS activity as they work through the steps?
* Are students aware of changes in scale as they zoom in and out on the map?

Concluding the lesson

- Engage students in a discussion about the observations and discoveries they made during their exploration of the U.S. map.
- Ask students about their impressions of different regions of the United States.
- Ask students to compare their experience working with a GIS map to their experience working with paper maps.
- Has this activity raised any questions that students would like to explore further?
- How can GIS help students to learn about U.S. history?
- Has this activity changed students' ideas about maps?

Extending the lesson

- Have students research landforms of America—mountains, deserts, plains, etc.
- Using layers from the GIS map, ask students to describe the relationship between rainfall and elevation. Have them give at least one reason for the relationship.
- Using layers from the GIS map, ask students to describe the relationship between rainfall and vegetation. Have them give at least one reason for the relationship.
- Using layers and information from the GIS map, ask students how much rainfall the grasslands receive. Mention that the grasslands later became the major crop-producing area of the United States. Ask students how so much wheat, corn and other grains can be grown on land that doesn't receive much rainfall?
- Have students research the Native Americans that lived in America before the Europeans came. What types of governments did they have? What types of rules? What languages did they speak? How were their societies organized? Did they have sporting events or competitions?

References

- http://countrystudies.us/united-states/geography-7.htm
- http://encarta.msn.com/encyclopedia_761589809/Westward_Movement_American.html
- http://www.nativeamericans.com/index.htm

M ● ● ● ○
L ● ● ○

Student activity answer key

Answers appear in blue.

Module 3, Lesson 1

Early settlement patterns of the United States

Before Europeans came to America, Native Americans had been living on the land for thousands of years.

Starting in the early 1600s, many people from Europe started coming to America to settle the same lands where millions of Native Americans had lived, hunted, fished, farmed, and traveled. The first European settlements were along the Atlantic Coast.

In the first 150 years after Europeans started coming to America (until about 1765), European settlement spread from the Atlantic Coast to the Appalachian Mountains. In the next one hundred years, settlement advanced all the way to the Pacific Ocean. As European settlements spread westward, Native American tribes were killed or driven off their lands and forced to move westward onto reservations.

This lesson is about the movement of European peoples from the settled regions of the eastern United States to lands farther west. As these people moved west, their settlements formed different patterns on the land. Sometimes the settlements formed a regular pattern; other times, the settlements were irregular, or scattered. Sometimes the settlements were close to transportation routes, such as rivers; other times, they were located close to each other, forming clusters.

Let's explore the historic settlement patterns we can still see among the cities of today.

Step 1: Start AEJEE.

1. Ask your teacher how to start the AEJEE software.

2. Click the **Maximize** button at the top of the AEJEE window. Now the AEJEE window fills your screen.

Step 2: Open the project.

1. Click the **Open** button.

2. Navigate to your **OurWorld1\Module3** folder.

3. Choose **SettlementPatterns.axl**.

4. Click **Open**.

   When the project opens, you see a map of the United States on your screen. (You don't see Alaska and Hawaii in this view.)

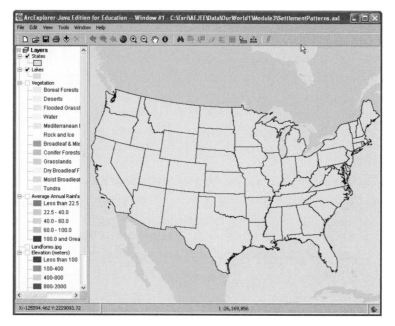

On the left side of the map, you see a list of layers. Layers are used to show geographic data on a GIS map. Each layer has a name and a legend. On the top, you see buttons and tools that you will use during this activity.

Take a minute to label the states.

5. Click the **States** layer name to highlight it. It becomes active.

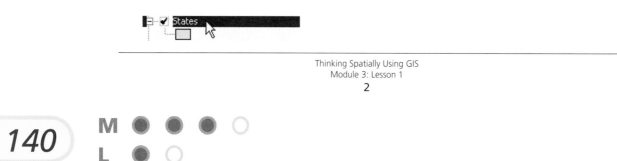

Making a layer active tells the GIS that this is the layer you want to work with.

6. Right-click the **States** layer name and choose **Properties**.

7. Click the **Labels** tab.

8. Under **Label features using**, choose **STATE_ABBR**. This field contains the abbreviations for the state names (e.g., California is CA)

9. Click **Effects**.

10. Click the box next to **Glow**.

11. Change the Glow color from Yellow to **Cyan**.

12. Click **OK**.

13. Click **OK** again.

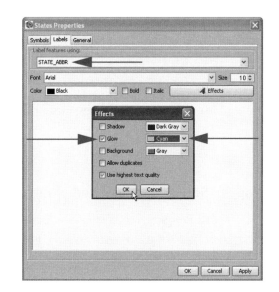

Now the states are labeled with their abbreviations, which have a glow around them. In order to see their full names, you will turn on MapTips.

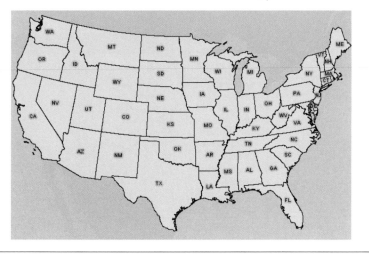

14. Click the **MapTips** button. The MapTips window opens.

15. Under **Layers**, click **States**.

16. Under **Fields**, click **STATE_NAME**.

17. Click **Set MapTips**.

18. Click **OK**.

19. Now when you hold your cursor over a state, the full name of the state appears.

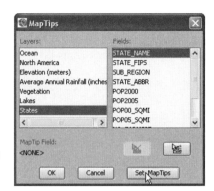

Q1 Look at the sizes of the states. What pattern do you see when you move from east to west? (Circle all the correct answers.)

   a. Eastern states are smaller.

   b. Eastern states are larger.

   c. Western states are smaller.

   d. Western states are larger.

Step 3: Settlements started along the East Coast.

The first European settlements developed along the East Coast. The British established their first settlement at Jamestown, Virginia, in 1607. Settlers established more towns in Plymouth and Boston, Massachusetts, and elsewhere between 1620 and 1630.

You will add cities to the map. Even though the cities you add are modern cities, they still reflect the settlement patterns that began with the early European settlements.

1. Click the **Add Data** button.

2. Navigate to your **OurWorld1\ Module3\Data** folder.

3. Click **Cities.shp**.

4. Click **OK**.

   Cities of the United States appear on the map. You'll change their color.

5. Right-click the **Cities** layer name and choose **Properties**.

6. Next to **Color**, click the drop-down arrow and choose **Green**.

7. Click the check mark next to **Remove Outline** to uncheck it.

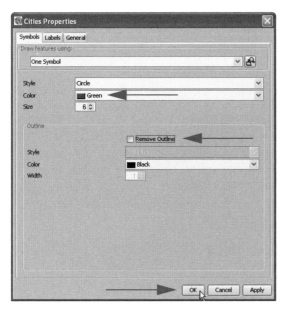

8. Click **OK**.

Now all the cities appear as green dots.

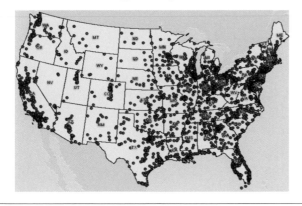

143

Q2 Which half of the country has more cities?

   a. Eastern half

   b. Western half

Q3 Look at the states along the East Coast. List three of the states that have cities along their eastern shore. (Hint: Use MapTips)

Massachusetts, Connecticut, New York, New Jersey, Maryland, Florida. Answers will vary.

Step 4: When European settlers moved west from the coast, they often followed waterways.

People used lakes and rivers to travel to new places to settle. People also used lakes and rivers to carry goods and travel from areas far from the ocean back to the shore, where most trading with Europe took place. This was especially true in the northeastern states.

1. Click the **Add Data** button.

2. Navigate to the **OurWorld1\Module3\Data** folder.

3. Click **NERivers.shp**.

4. Click **OK**.

Rivers of the northeast appear on the map. You will zoom in for a closer look.

5. Click the **NERivers** layer name to highlight it. It becomes active.

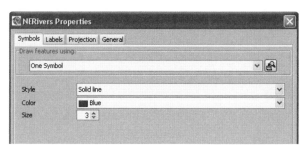 6. Click the **Zoom to Active Layer** button.

The map zooms to the northeast part of the country. The rivers are hard to see, so you will change their symbol.

7. Right-click the **NERivers** layer name and choose **Properties**.

8. Click the drop-down arrow next to **Color** and choose **Blue**.

9. Increase the **Size** to **3**.

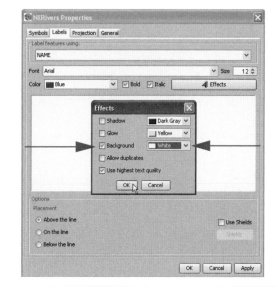

10. Click the **Labels** tab.

11. Under **Label features using** choose **NAME**.

12. For **Color**, choose **Blue**.

13. Increase the **Size** to **12**.

14. Click the box next to **Bold**.

15. Click the box next to **Italic**.

16. Click the **Effects** button.

17. In the Effects window, click the box next to **Background**.

18. For the Background **Color**, choose **White**.

19. Click **OK**.

20. Click **OK** again.

Now the northeast rivers appear as thick blue lines with blue labels on white backgrounds.

145

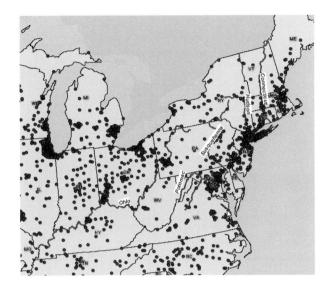

The British people used the waterways as highways, just like the Native Americans did. Some settled along the coast of Chesapeake Bay. It was here that the Native Americans introduced the settlers to tobacco.

21. To find Chesapeake Bay, look at your map and follow the clues below.

22. Clue #1: Find the Susquehanna River. (Hint: It has a blue label.)

23. Clue #2: Find the place where the Susquehanna River empties into the ocean.

Now you have found Chesapeake Bay.

Q4 **Are there cities along the coast of Chesapeake Bay? (Circle the correct answer.)**
  a. Yes

  b. No

Q5 **Most of these cities are in what state? (Hint: Use MapTips.)** Maryland

The Dutch settled along the Hudson River. Henry Hudson, an Englishman working for a Dutch trading company, was the first European to discover the Hudson River Valley. He was sailing along America's north Atlantic Coast when he found it by accident in 1609. He entered what is now New York Bay and sailed up the river that was later named for him. His explorations led to the area being settled by the Dutch between 1609 and 1664. The Dutch settlement was first known as New Amsterdam, but was later renamed by the British to New York.

M ● ● ● ○
L ● ○

24. Turn off the **Cities** layer by clicking the check mark next to its name.

25. Look at the map and find the Hudson River. (Hint: It has a blue label.)

26. Find the place where the Hudson River empties into the ocean.

27. Turn the **Cities** layer back on.

Q6 Are cities found along the Hudson River where it empties into the ocean? (Circle the correct answer.)

    a. Yes

    b. No

Q7 Most of these cities are in what state? (Hint: Use MapTips.) New Jersey

Step 5: By about 1750, European settlements had moved west as far as the Appalachian Mountains.

The early European settlers wanted to stay close to each other and to their home countries across the Atlantic. The rugged Appalachian Mountains discouraged them from moving farther west for many years because it was difficult to cross, with few openings and few trails for wagons.

1. Right-click the **States** layer name and choose **Properties**.

2. For **Style**, choose **Transparent fill**.

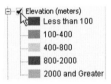

3. Click **OK**.

Now you see just the outlines of the states with their labels.

4. Turn on the **Elevation (meters)** layer by clicking the box next to the Elevation layer name. (Be patient while this layer draws.)

The elevation layer shows high areas (mountains) and low areas (plains).

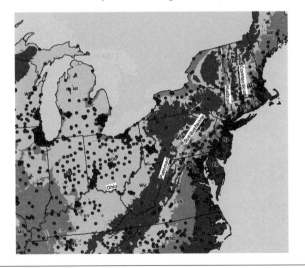

Q8 Look at the map and the Elevation legend. Which color do you think represents the Appalachian Mountains? (Circle the correct answer.)

   a. **Dark green**

   b. **Light green**

   c. **Light brown**

   d. Dark brown

   e. **Pink**

Q9 How high are the Appalachian Mountains in meters? (Circle the correct answer.)

   a. **Less than 100**

   b. **100–400**

   c. **400–800**

   d. 800–2,000

Hint: A meter is a little more than 3 feet (3.28 feet), so if you multiply the number of meters by 3, you will know approximately how high the mountains are in feet.

5. To see the mountains better, turn on the **Landforms.jpg** layer by clicking the box next to its name.

This layer is an image (or picture) that shows the mountains and plains. It is partly see-through, so you can see the elevation layer at the same time. Now you can really see where the mountains are. You'll zoom out for a better view.

6. Click the **Zoom Out** tool.

7. Click the word **Ohio** near the center of the map. (This is the blue label for the Ohio River.)

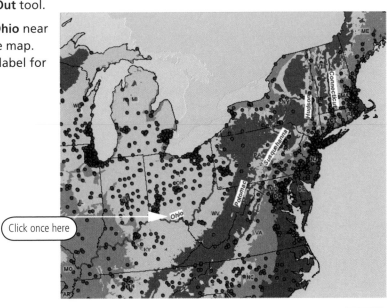

149

The map zooms out so you can see more of the Appalachian Mountains.

> **Note:** If you make a mistake, click the **NERivers** layer name to highlight it and click the **Zoom to Active Layer** button. Then click the **Zoom Out** tool and try again.

Step 6: When Europeans finally moved west of the Appalachian Mountains in about 1770, they quickly settled along waterways, such as the Ohio River.

Native Americans who lived west of the Appalachians wanted to keep their land and their way of life. They fought a bloody war with the British. Then the British government passed a law banning European settlers from moving west of the Appalachians.

However, this law was almost impossible to enforce because the British government was so far away across the Atlantic Ocean. In the 1770s, Richard Henderson, who hoped to make money in the land business, got permission from the local Native Americans (Cherokees) to settle in Kentucky. He hired the famous frontiersman Daniel Boone to lead families across the Appalachians at a place called the Cumberland Gap. They founded the town of Boonesboro in central Kentucky.

1. Turn off the **Landforms.jpg** layer by clicking the check mark next to its name.
2. Click the **Previous Extent** button to return to a closer view.
3. Look at the Ohio River on the map.

Q10 Which of these patterns most closely matches the pattern of cities along the Ohio River? (Circle the correct answer.)

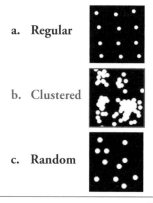

a. **Regular**

b. **Clustered**

c. **Random**

M ● ● ● ○
L ● ○ ○

Step 7: Settlement continued to move west as far as Kansas and Nebraska.

After the Americans defeated the British in the American Revolution (1775–1783), the U.S. government encouraged people to move west by selling them land. The government paid war veterans by giving them free land.

European settlers brought a different kind of agriculture to the land. While Native Americans had grown several crops on a single piece of land, European settlers used one piece of land to grow a single crop. They also tried to grow more than they could eat, so they could sell the excess to people in the east.

The new approach to farming exhausted the soil more quickly than Native American farming, and it created a demand for more and more land. This helped push settlement farther westward.

1. Click the **Zoom Out** tool.

2. Place your mouse over the western end of the Ohio River and click once.

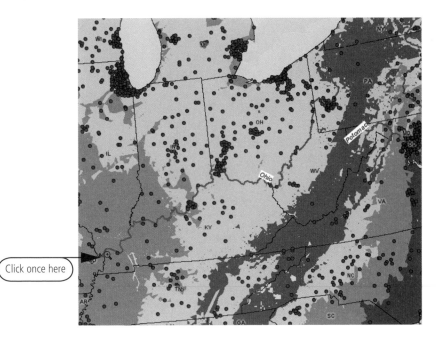

Click once here

> **Note:** If you make a mistake, click the **NERivers** layer to highlight it, and click the **Zoom to Active Layer** button. Then click the **Zoom Out** tool, and try again.

151

Q11 Look at the map and the Elevation legend. What happens to elevation as you move west into Kansas, Nebraska, and South Dakota?

a. Elevation increases (elevation is higher)

b. Elevation decreases (elevation is lower)

c. Elevation stays the same

Q12 Look at the elevation layer on the map and compare it with the cities layer. Can you explain how elevation affects where people live? Fewer cities are located at high elevations. More cities are located at low elevations, especially along coastlines.

3. Turn off the **Elevation** layer by clicking the check mark next to its name.

4. Turn on the **Average Annual Rainfall** layer. (Be patient while this layer draws.)

This layer shows the average amount of rain (in inches) that falls in a year.

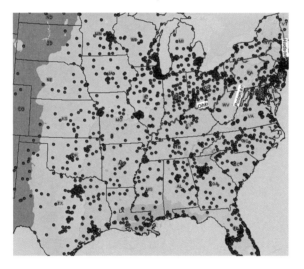

Q13 Look at the legend for the Average Annual Rainfall layer in the table of contents. Which color best represents the amount of rainfall in the states of Oklahoma, Kansas, and Nebraska? (Circle the correct answer.)

a. Red

b. Orange

c. Green

d. Blue

Q14 How much average annual rainfall does this color represent? (Circle the correct answer.)

    a. **Less than 22.5 inches**

    b. 22.5–40 inches

    c. 40–60 inches

Q15 Look at the rainfall layer and compare it with the cities layer. Can you explain how rainfall affects where people live? There are fewer cities in areas that have low rainfall (red areas and orange areas that are close to red areas).

Step 8: By the 1840s, settlement had moved west of the Mississippi River onto grasslands in the center (interior) of the country.

1.   Turn off **Average Annual Rainfall** by clicking the check mark next to its name.

2.   Turn on the **Vegetation** layer.

The Vegetation layer shows the type of plant growth in an area. (Example: Grasslands)

3.   Click the **View** menu and choose **Pan Panel**.

A frame appears around your map. Each side of the frame has a small white arrow.

4.   Click **once** on the small white arrow on the left side of the frame to **Pan West**.

153

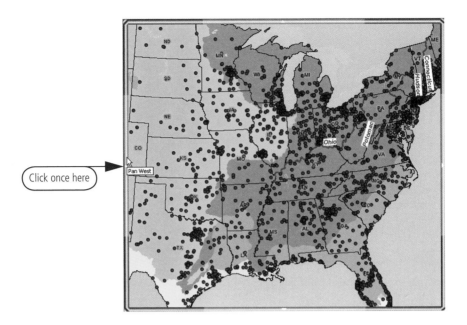

The map moves to show more of the West.

Q16 Look at the Vegetation legend. What color represents grasslands? (Circle the correct answer.)

a. Green

b. Pink

c. Blue

d. Orange

Step 9: **The Mississippi River and the rivers that flowed into it (tributaries) offered easy routes to the central (interior) grasslands.**

Once again, settlers were drawn westward in hopes of making a better living and improving their lives. West of the Mississippi, European settlers could find more ways to make a living than just farming. There were minerals like gold and silver for mining, forests for harvesting trees (in the Far West), and vast lands for raising crops and grazing cattle.

1. Turn off **NERivers** by clicking the check mark next to the layer name.

2. Click the **Add Data** button.

3. Navigate to your **OurWorld1\Module3\Data** folder.

4. Click **Rivers.shp**.

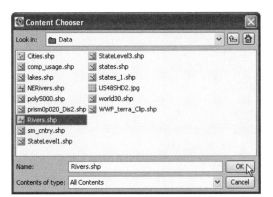

5. Click **OK**.

6. Right-click the **Rivers** layer name and choose **Properties**.

7. Change the **Color** to **Blue**.

8. Increase the **Size** to **2**.

9. Click **OK**.

   You see rivers of the central states. You will select the Mississippi River and its tributaries.

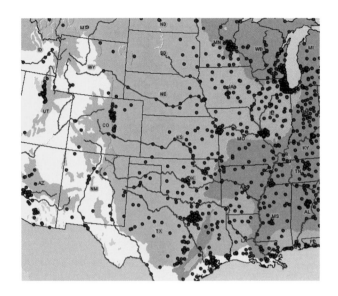

10. Click the **Rivers** layer name to highlight it.

11. Click the **Query Builder** button. The Query Builder window opens.

12. Under **Select a field**, click **SYSTEM**.

13. Click the **= (equals)** button.

14. Under **Values**, click **Mississippi**.

    Your query should look like this:
    (SYSTEM = 'Mississippi')

15. Click the **Execute** button.

Q17 **Look at the Query Results at the bottom of the Query Builder window. How many rivers make up the Mississippi River system? 13**

16. Close the Query Builder window.

The Mississippi River and its tributaries are highlighted in yellow on the map.

M ● ● ● ○
L ● ○

Q18 Look at the grasslands area on your map. As you move west across the grasslands, the pattern of cities changes. Describe these changes in your own words. Look back at Q10 and see if you can match any of the patterns.

Answers may include the following:

As you move from east to west across the grasslands, you may notice the following:

- The number of cities decreases.

- Many of the cities are located along rivers of the Mississippi system.

- The pattern of cities is random or clustered.

Step 10: Settlement was scattered from the Rocky Mountains westward.

By the 1840s, settlement had moved only a few hundred miles west of the Mississippi River.

Between 1841 and the late 1860s, westward settlement jumped across the middle of the country to Oregon and California on the Pacific Coast.

> In January 1848, gold was discovered in California. This discovery sparked the gold rush of 1849.
> - Thousands of people traveled to California using overland trails.
> - Just as many people came by ship.
> - Over the next few years, hundreds of thousands more people crossed the continent for gold.
> - Few people got rich, and many returned home.
> - Enough stayed to establish California as a state in 1850.

1. Click the small white arrow along the left border of the map to **Pan West**.

   Now you can see all the way to the West Coast.

   Click once here

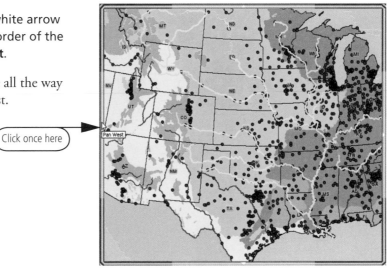

157

2. Click the **Pan** tool.

3. Place the cursor (which looks like a hand) over the UT label for Utah.

4. Click and *hold* down the mouse button.

5. Move the hand down until you can see all the western states.

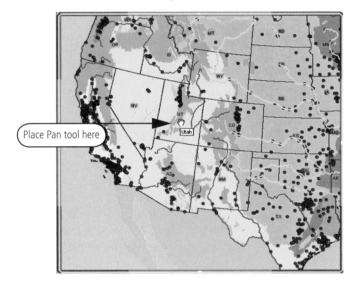

Place Pan tool here

Q19 Look at the map and look at the Vegetation legend. List three western states that are mostly covered with deserts: Answers may include the following:

Nevada

Utah

Arizona

New Mexico

Wyoming

Idaho

Q20 Which pattern most closely matches the pattern of cities in these states? (Circle the correct answer.)

a. Regular

b. Clustered

c. Random

6. Turn off **Vegetation** by clicking the check mark next to the layer name.

7. Turn off **Rivers**.

8. Turn off **Cities**.

9. Turn on **Average Annual Rainfall** by clicking the box next to the layer name.

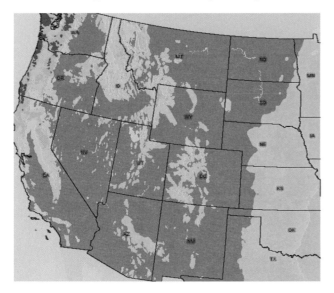

 Q21 Look at the map, and look at the Average Annual Rainfall legend. How much rain falls each year in most of the western states? (Circle the correct answer.)

   a. Less than 22.5 inches

   b. 22.5–40 inches

   c. 40–60 inches

So far, you have learned that some of the western states are deserts and that most of the western states have little rainfall. Let's look at elevation next.

10. Turn off **Average Annual Rainfall** by clicking the check mark next to the layer name.

11. Turn on **Elevation**.

Q22 Look at the map, and look at the Elevation legend. What is the elevation of most of the western states? (Circle the correct answer.)

   a. 400–800 meters

   b. 800–2,000 meters

   c. 2,000 meters and greater

12. Turn on the **Landforms.jpg** layer. After the map redraws, you see landforms (mountains, valleys, and plains) and elevation at the same time.

13. Turn on the **Cities** layer again.

(Q23) Look at the elevation layer and the landforms layer on the map. You see mountains in high elevation areas. Which pattern most closely matches the pattern of cities in these areas? (Circle the correct answer.)

a. **Regular**

b. **Clustered**

c. **Random**

Answer: c is the preferred answer; b is also acceptable

14. Click the **Lakes** layer name to highlight it. It becomes active.

15. Click the **Zoom to Active Layer** button.

The map zooms out.

M ●●●○

L ●○

Q24 **Notice how the number and pattern of cities change as you move west across the United States. Describe these changes in your own words. Look back at Q23 and see if you can match any of the patterns.** Answers may include the following:

As you move from east to west across the United States, you may notice the following:

- The number of cities decreases.

- Cities are not as close together.

- The pattern of cities is more random.

Q25 **List two reasons why settlement patterns are different in the West than in the East.** Answers may include the following:

- Much of the West has high elevation and mountains.

- Much of the West has little rainfall.

- Much of the West is covered with deserts.

- There are fewer rivers in the West.

Step 11: Save your work and exit AEJEE.

1. Ask your teacher how to save your work.

2. Click the **File** menu and choose **Exit**.

Conclusion

After the arrival of Europeans, the settlement patterns on lands that became the United States began with people moving from east to west. People moved westward as farmland increasingly became available. Then during the Industrial Revolution (1820–1870), people moved from farmlands to cities to find jobs. Once most of the jobs were in cities, people began moving from city to city. Since the 1970s, people have been moving away from cities for a variety of reasons—to places like the South, where the population had not increased for a long time.

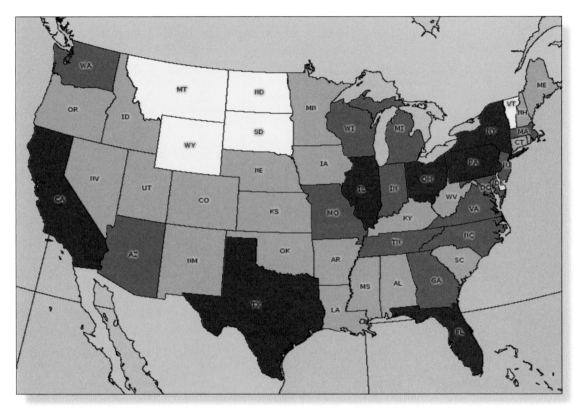

A graduated color map showing total U.S. population (the yellow states have the lowest population, and the blue states have the highest population) helps students identify and interpret population patterns.

Module 3: Lesson 2

Patterns of a growing population

Lesson overview

Students will explore the role of the United States Census. Using data obtained in the 2000 Census, they will examine population patterns across the United States. Students will profile their state and examine data on population, population density, and other interesting statistics on mobility and Internet access. They will use GIS to examine and interact with a number of graduated color maps and query data to answer relevant questions.

Estimated time

Approximately 60 minutes

Materials

The student activity and worksheet can be found in the student workbook or on the Software and Data CD. Install the teacher resources on your computer to access them.

Location: OurWorld_teacher\Module3\Lesson2
* Student activity and worksheet: M3L2_student.pdf

Objectives

After completing this activity the student will be able to do the following:

* Explain population patterns across the United States
* Profile the information for their state
* Interpret thematic maps to answer questions and draw conclusions
* Work with tabular data
* Query data to find answers and make decisions

GIS tools and functions

☐ Set MapTips to display information about features without clicking

☐ Identify a feature on the map

☐ Use the Query Builder to select data based on certain criteria

☐ Zoom to the active layer

☐ Erase any selection made on the map features

- Turn layers on and off
- Activate a layer
- Query attribute data
- Obtain statistics for data
- Label features
- Sort attributes
- Select features

National geography standard

Standard	K-4	5-8
3 How to analyze the spatial organization of people, places, and environments on Earth's surface	That places and features are distributed spatially across the Earth's surface	How to use the elements of space to describe spatial patterns
9 The characteristics, distribution, and migration of human populations on Earth's surface	The spatial distribution of population	The demographic structure of a population
10 The characteristics, distributions, and complexity of Earth's cultural mosaics	How the characteristics of culture affect the ways in which people live	The spatial distribution of culture at different scales (local to global)
18 To apply geography to interpret the present and plan for the future	The spatial dimensions of social and environmental problems	How to apply the geographic point of view to solve social and environmental problems by making geographically informed decisions

M ● ● ● ○
L ● ●

Teaching the lesson

Introducing the lesson

Begin this lesson by reviewing these concepts:

- The difference between total population and population density
- Spatial patterns: cluster, random, and uniform
- How the U.S. Census can help us study the history of the United States
- The use of statistics in geography and mapping
- How census data is simplified so that it can be studied more easily
- The unique characteristics of each region of the United States
- Thematic mapping using graduated color maps
- The individual characteristics of your state and the people who live there

The student activity

We recommend you complete this lesson before completing it with students. This will allow you to modify the instructions to accommodate the specific needs of your students.

Teacher notes

- Explain the lesson to the students, and make sure they know where to answer the questions.
- For younger grades, you can conduct the GIS activity as a teacher-led activity and have students follow along. You can lead students through the GIS steps and ask them the associated questions as a class.
- Students will each need a printed copy of the activity so they can answer questions throughout. They can mark their answers directly on the activity sheet. Alternatively, you can provide a separate answer sheet.
- Ideally, students will have access to their own computers. However, students can complete the lesson in small groups.
- Some questions do require classroom and/or group interaction. You can decide on the best way to handle these questions.
- We recommend that students save their work as they progress through the GIS activity. Students can use either the Save command (to save their changes to the original map) or the Save As command (to save their changes to a new map). Please explain to students where and how they should save their work.
- Each student will complete a worksheet while going through the activity.
- To explain population density, have the students calculate the density of the students in your classroom (number of students / area of classroom = density).
- When you open the AXL file, the software opens at its standard size. The list of layers can be widened by dragging the line that separates the list from the map to the right. You might need to do this for the students to see all the numbers in the legends.
- Software notes:
  - Students should maximize the software window *before* opening an AXL file.
  - If students close AEJEE before completing the GIS activity, the MapTips function is not saved. Students will have to set MapTips again when they reopen the map.

The following are things to look for while the students are working on this set of activities:

- Are students thinking about the underlying geographic concepts as they work through the steps (such as the population of their community and how population affects decision making)?
- Are students answering the questions as they work through the steps?
- Are students experiencing any difficulties with the buttons, tools, and mouse clicking, etc.?

Concluding the lesson

- Engage students in a discussion about the United States and its population. For example, you can discuss issues such as how fast the population is growing, where people are living, and how population patterns might change in the future.
- Have students do Internet research to learn more about the future growth of the United States and/or world population and to learn about the next U.S. Census.
- Ask students what they know about the populations in other countries.
- Discuss the value of accurately counting people. For example, how would this affect the location of schools and construction of new homes?
- Talk about how your community compares with the state and the country in terms of population and other characteristics. Does their community represent the state as a whole?
- Has this activity raised any questions that they would like to explore further?
- How can GIS help them to learn about people and places? How can GIS help them organize information and find answers to questions?

Extending the lesson

- Download a 2000 Census form and have each student complete the form (short or long form). Do a mini census of the classroom.
- Examine other kinds of questions that could be asked on the U.S. Census.
- Choose one state from each region of the United States to profile, and see how those states compare to yours.
- Discuss population patterns in the country and how they relate to history, landforms, climate, etc.

References

- www.census.gov
- http://people.howstuffworks.com/census1.htm
- http://www.ourtownamerica.com/sponsor/movingstats.php

Student activity answer key

Answers appear in blue.

Module 3, Lesson 2

Patterns of a growing population

The United States population has grown quickly during the past several hundred years. Keeping track of the nation's population dates to the country's origins. The U.S. Constitution adopted in 1787 called for a population count every 10 years, starting in 1790. This process, called the census, would keep track of the population, its activities, and its movements. More importantly, the census would ensure that each state received fair and accurate representation in the U.S. House of Representatives.

The 1790 Census recorded almost 4 million people. By comparison, the 2000 Census counted almost 300 million. That's more than 70 times the number of people that lived in the United States 210 years ago! It is estimated that by 2050 there will be 392 million people living in the United States! The United States now is the third most populated country in the world after China and India.

So, what kind of questions is the U.S. Census Bureau asking? The census used to ask simple questions, such as how many people lived in a home and how many cows did a family own. But the census increasingly has become a way to understand where and how government spends money in such areas as education, health care, and housing. Today, the census asks questions about many aspects of life in the United States. These questions range from wanting to know how much crime we have to how much grain, fruits, and vegetables we produce—and where we produce them.

Let's explore some of the characteristics of the more than 300 million people who live in the United States today! We will also examine how your state compares to the rest of the country. First, let's learn a little bit about where you live.

Q1 What is the name of your state? Write it on your "Patterns of a growing population" worksheet.

A lot of decisions are made based on the information collected in the census. Let's look at some of the patterns that showed up based on how we answered questions during the 2000 U.S. Census.

Step 1: Start AEJEE.

1. Ask your teacher how to start the AEJEE software.

2. Click the **Maximize** button at the top of the window. Now the AEJEE window fills your screen.

Step 2: Open the project.

1. Click the **Open** button.

2. Navigate to your **OurWorld1\Module3** folder.

3. Choose **USCensus.axl**.

4. Click **Open**.

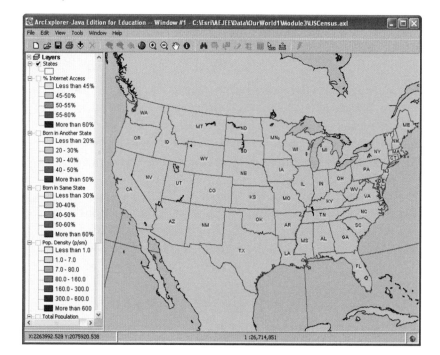

A map of the United States will be on your screen along with other background information. A list of layers is on the left side of the map. Layers are used to show geographic data on a GIS map. Each layer has a name and a legend. You can turn each layer on and off. At the top of the map, you see buttons and tools that you will use during this activity.

Step 3: Where everyone lives.

As mentioned, one of the reasons for the census is to see where everyone in the country lives. During the first census, marshals rode on horseback across the country and collected information by talking with people personally. At that time, most of the population lived on the East Coast in the original 13 states.

Things are much different today.

1. Click the name of the **States** layer to make it active. It will become highlighted.

2. Click the **MapTips** button. This will open the MapTips window.

3. Under **Layers**, click **States** (you might have to scroll down).

4. Under **Fields**, click **STATE_NAME**.

5. Click **Set MapTips**.

6. Click **OK**. Now when you hold your cursor over a state (do not click), its entire name will appear.

> **Note:** If you live in Alaska or Hawaii, click the **Zoom to Active Layer** button to view your state. You can leave the map in this position for the remainder of the lesson. To go back to a view of the 48 states, click the **Zoom to Previous Extent** button.

7. Scroll down the list of layers. Turn on the **Total Population** layer by clicking the box next to its name.

169

8. Click once on the **Total Population** layer name to make it active. It will become highlighted.

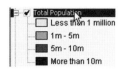

This is a *graduated color* map showing total population for the country. States with darker colors have higher populations. States with lighter colors have lower populations.

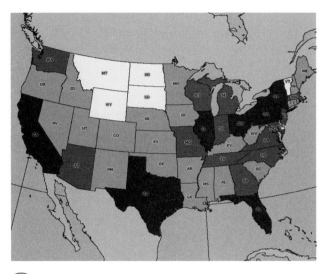

 Q2   **Look at your map and find the darkest colored states. From those states, select three that each have populations of more than ten million people.** Answers may include the following:

1. California

2. Texas

3. Florida

4. New York

5. Pennsylvania

6. Ohio

7. Illinois

Q3   **Look at the location of all the states that fall in the highest population category. What do they have in common? (Circle the correct answer.)**

a. **They are all next to one another.**

b. **They all start with the letter C.**

   c. They all have some of their border along water.

   d. They are all in the middle of the country.

Q4 **List three of the states that have fewer than one million people (the lightest color)? Answers may include the following:**

   1. Montana

   2. North Dakota

   3. South Dakota

   4. Wyoming

   5. Vermont

   6. Delaware

   7. Alaska

Q5 **What pattern do these states show?**

   a. **They are mostly clustered together.**

   b. **They are randomly spread apart.**

   c. **They are evenly distributed across the country.**

9. Right-click the **Total Population** layer and choose **Attribute Table**. The attribute table for this layer will appear.

10. Scroll to the right and find the **POP2000 (population in the year 2000)** field.

11. Right-click the **POP2000** field and choose **Sort Descending**. This will rearrange the records in order from the highest population to the lowest population.

Q6 Scroll back to the left to see the state names. Which state has the highest population (the one at the top of the list)? California

12. Right-click the **POP2000** field and choose **Sort Ascending**. This will rearrange the records in order from the lowest population to the highest population.

STATE_FIPS	STATE_ABBR	POP2000	Region
15	HI	1211537	West
53	WA	5894121	Sort Ascending
30	MT	902195	Sort Descending
23	ME	1274923	Sort Selected Data to Top
38	ND	642200	Export Selected Data

Attributes of Total Population

Q7 Scroll back to see the state names. Which state has the lowest population (the one at the top of the list)? Wyoming

13. Close the **Attribute Table**.

Let's see where your state fits in.

 14. Click the **Find** tool.

15. Under **Value**, type the name of your state. You must type it with a capital letter at the begining and make sure you have the correct spelling.

16. Under **Layers to Search**, click **Total Population**.

17. Click **Find**. Your state will appear in the list on the right.

18. Click **Select**.

19. Click **Zoom To**. This will zoom you right into your state.

> **Note:** If your state does not appear, check your spelling and make sure the first letter is capitalized.

20. Close the **Find** window. You will be zoomed in to your state, and it is highlighted in yellow.

21. Click the **Identify** tool.

Module 3: Lesson 2

22. Click your state.

23. In the window that appears, locate the **POP2000** field.

Field	Value
ObjectID	24
STATE_NAME	California
STATE_FIPS	06
STATE_ABBR	CA
POP2000	33871648
Region	West
SUB_REGI_1	Pacific
URBAN	31989663
RURAL	1881985
MALE	16874892

Identify Results — 1 feature — California — Layer: Total Population

Q8 Write the population of your state in the appropriate location on your worksheet.

24. Close the **Identify Results** window.

25. Click the **Previous Extent** button. This will zoom you out to the continental United States.

Step 4: More people, less space.

In the previous step, you saw a map that showed the distribution of the total U.S. population. Another way that analysts look at population is to look at *density*.

Population density is the number of people in a particular area or unit. Usually, this is shown in terms of the number of people per square mile. In Los Angeles County, California, about 2,300 people live in each square mile. In Yellowstone County, Montana, about 49 people live in each square mile. Therefore, Los Angeles has roughly 47 times more people per square mile!

1. Turn off the **Total Population** layer by clicking the check mark next to its name.

2. Turn on the **Pop. Density (p/sm)** layer by clicking the box next to its name.

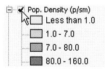

Pop. Density (p/sm)
- Less than 1.0
- 1.0 - 7.0
- 7.0 - 80.0
- 80.0 - 160.0

Let's find the states with the highest and lowest population densities.

Module 3: Lesson 2

173

3. Click the **Pop. Density (p/sm)** layer name to make it active. It will become highlighted.

4. Click the **Zoom to Active Layer** button. This will zoom the map out to the entire country, including Alaska and Hawaii. (Hint: If you were already zoomed out, then you do not need to do this again.)

5. Right-click the **Pop. Density (p/sm)** layer name.

6. Choose **Attribute Table**.

7. Scroll to the right until you find the **Pop_Dens** (population density) field.

8. Right-click the **Pop_Dens** field and choose **Sort Ascending**. This will rearrange the records in the table from lowest to highest population density.

9. Click the first record in the list. It will be highlighted in blue.

Q9   **What is the name of the state that has the lowest population density?**
State: Alaska

10. Close the **Attribute Table**.

Q10 **Can you think of two reasons why this state has the lowest population density?**
Answers may include the following:

1. The weather is cold.

2. It is very far from the contiguous states (states that are connected).

3. The land, or terrain, is rugged.

4. The types of industries in Alaska are more limited than in many other states.

5. The actual area of the state is quite large, so people can spread out more.

 11. Click the **Previous Extent** button. This will bring you back to a map of the continental United States. (If you live in Alaska or Hawaii, do not click **Previous Extent**.)

12. Right-click **Pop. Density (p/sm)** again.

13. Choose **Attribute Table**.

14. Scroll to the right and locate the **Pop_Dens** field.

15. Right-click the **Pop_Dens** field and choose **Sort Descending**. This will rearrange the list from highest to lowest population density.

Attributes of Pop. Density (p/sm)		
PCTNOKIT	Pop_Dens	Area
1.82	173.38	6007.02
1.15	86.63	Sort Ascending
2.69	6.07	Sort Descending
3.68	39.57	Sort Selected Data to Top
1.98	8.97	Export Selected Data

16. Click the first record. It will be highlighted in blue in the table and in yellow on the map.

Q11 **Scroll back to see the names of the states. Which state has the highest population density?**
State: New Jersey

Q12 **Can you think of two reasons why this state has the highest population density?**
Answers may include the following:

1. It is a small state, so there is not a lot of space for people.

2. There are lots of jobs nearby (e.g., New York City).

3. The weather is moderate.

4. People have access to transportation.

5. It is near major cities.

175

17. Close the **Attribute Table**.

18. Click the **Clear All Selections** button.

Q13 Look at your map and notice how population density is spread across the states. Describe any patterns you see. Use words such as "located together" (or clustered), "without any pattern" (or random), and "looking the same as the others" (or uniform). Answer: Students can answer this creatively. Appropriateness will vary depending on grade level. The following are some points for students to consider:

- Other than California, Washington, and Florida, the states with higher density are clustered in the Northeast.

- The states with mid-density are essentially clustered in the middle of the country.

Let's see where your state fits in.

19. Click the **Identify** tool.

20. Click your state. (Hint: If it's hard to find your state, hold your cursor over the state until the name pops up in the MapTip. Then click your state.)

21. In the window that appears, scroll down and locate the **Pop_Dens** field.

Q14 What was the population density (Pop_Dens) of your state at the 2000 Census? Write it on your worksheet.

22. Close the **Identify Results** window.

Step 5: Moving around the country.

People who live in the United States are on the move. Did you know that Americans on average move eleven times in their lives? While some people never move, about one in five people move each year—that's 20 percent of the population. Some move for family reasons such as getting married. Others move to start new jobs, go away to college, or buy a different house.

It's not easy keeping track of population movement. The U.S. Census asks people their current address but also asks if they have moved in the last five years.

176

M
L

1. Click the **Previous Extent** button. You will go back to a view of the continental United States. (Hint: If you are already at this view, then you do not need to click the button.)

2. Turn off the **Pop. Density (p/sm)** layer.

3. Turn on the **Born in Same State** layer. This layer shows the percent of people that were born in the same state they currently live in.

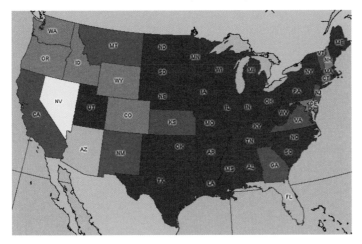

Let's look at the states in the highest category.

4. Click the **Born in Same State** layer name to highlight it. It becomes active.

5. Click the **Query Builder** button.

6. Under **Select a field**, choose **PCTINSTATE**.

7. Click the >= (greater than or equal to) button.

8. Put your cursor next to the >= sign and type **70**.

9. Your query should look like this: **(PCTINSTATE >= 70)**

10. Make sure the parentheses are there!

11. Click **Execute**. The states that have more than 70 percent of people who are born in the state will appear in the list at the bottom of the window. They will also appear in yellow on your map.

Module 3: Lesson 2

177

Q15 Select any five states from those that become selected and write their names here: Answers may include the following:

1. North Dakota
2. Wisconsin
3. Minnesota
4. Iowa
5. Pennsylvania
6. Ohio
7. West Virginia
8. Kentucky
9. Alabama
10. Mississippi
11. Louisiana
12. Michigan

12. Close the **Query Builder**. The states will still be highlighted.

Q16 In which part of the country are most of these states located?

a. East

b. West

Q17 Discuss with your classmates why you think people want to remain living in these states. Answer: This is just a conversation the students should have in order to answer the following question.

Q18 List two reasons why people would not move away from the state they were born in. Answer: The following answers, as well as any other valid reasons, are acceptable.

1. Their family lives there.
2. They have good jobs.
3. They have a house.
4. They cannot afford to move.
5. They like where they live.
6. They like the weather.

Module 3: Lesson 2

7. Their friends live there.

8. They have never left the state.

(Q19) **Ask your classmates if they were born in the state they live in. How many said yes? How many said no?** Answers will vary.

 13. Click the **Clear All Selections** button.

Let's see where your state fits in.

 14. Click the **Identify** tool.

15. Find your state on the map.

16. Click your state. The **Identify Results** window appears.

17. Scroll down until you see the **PCTINSTATE** field.

(Q20) **What is the percent of people living in the state that were born there? Write this on your worksheet.** Answers will vary.

18. Close the **Identify Results** window.

19. Turn off the **Born in Same State** layer.

20. Turn on the **Born in Another State** layer.

This map shows the percent of people in each state who were born in a different state.

(Q21) **Did you notice a significant change in the map? (Hint: Turn the Born in Another State layer on and off to take another look.)**

a. Yes

b. No

(Q22) **Describe the change in pattern that you noticed.** The darker colors moved to the West of the country, showing that people in the west have typically moved there from other states.

21. Turn off the **Born in Same State** layer.

Look at the map. Which states have the highest number of people who were born in another state (the darkest color)? Write two of them here. Answers may include the following:

1. Wyoming

2. New Hampshire

3. Alaska

4. Nevada

5. Arizona

 What do you think makes these states so popular? Answers may include the following:

- Nicer weather (not in all cases)

- Better job opportunities

- Lower cost of living/less expensive housing

- Better lifestyle

- Attend college/university

22. In the list of layers, find the **Born in Another State** layer.

23. Click the colored box that represents the *lowest* number of people who were born in another state (Less than 20%). All the states that fall in that category will be highlighted in green on the map.

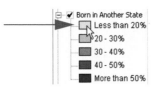

Q25 **Look at the map. Which states have the lowest number of people who were born in another state (the lightest color)? List three of them here:** Answers may include the following:

1. Massachusetts

2. New York

3. Pennsylvania

4. Illinois

5. Louisiana

6. Michigan

Q26 Why do you think so few people from other states move there? Answers may include the following:

- High cost of living

- Very large cities

- More pollution

- Colder weather (not in all cases)

- A lot of people live in cities (urban areas)

24. Click the **Born in Another State** layer name to highlight it. It becomes active.

25. Click the **Clear All Selection** button.

   Let's see where your state fits in.

26. Click the **Identify** tool.

27. Find your state on the map.

28. Click your state. The **Identify Results** window appears.

29. Scroll down until you see the **PCTOTHSTAT** field.

Q27 What is the percent of people living in your state who were born in another state? Write this on your worksheet.

30. Close the **Identify Results** window.

Step 6: Surfing the Internet.

Each year, the U.S. Census performs a survey called the American Community Survey. The survey collects interesting facts on the U.S. population and publishes them on the Web.

One of the pieces of data the survey examines is people's access to the Internet. The U.S. Census collects this information for each household in the United States.

1. Turn off the **Born in Another State** layer.

2. Turn on the **%Internet Access** layer by clicking the box next to the layer name.

Q28 **Look at your map. List three of the states where more than 60 percent of households have access to the Internet? (These will be the states in the darkest colors.)** Answers may include the following:

1. Alaska

2. Washington

3. Minnesota

4. Oregon

5. New Hampshire

6. Connecticut

7. New Jersey

8. Utah

9. Colorado

10. Virginia

Q29 **Look at your map. What pattern do you notice among the states where more than 60 percent of the population has Internet access? (Circle all that apply.)**

a. There is a random pattern—these states are spread all over the country.

b. There is a cluster pattern—the states are grouped together in certain areas.

c. There is a uniform pattern—the states are spaced evenly apart.

Answer: a or b—it will depend on the student's spatial perspective.

3. Click the **%Internet Access** layer name to highlight it. It becomes active.

4. Right-click the **%Internet Access** layer and choose **Attribute Table**.

5. Right-click the **INTACC_PCT** field. This is the percent of households with access to the Internet.

6. Choose **Sort Ascending**. This puts the states in order from lowest to highest percentage.

7. Hold down your **CTRL** key on the keyboard, and click the first three states in the table. They become highlighted in blue. They will also become highlighted on your map.

Attributes of % Internet Access			
COMP_TOT	INTACC_TOT	INTACC_PCT	COMACC_PC
537	435	39.55	48.82
544	462	42.31	49.82
935	785	44.1	52.53
390	323	44.86	54.17
996	846	45.8	53.93
899	749	45.95	55.15
409	354	47.39	54.75
787	687	48.31	55.34
1357	1171	49.02	56.8
951	813	49.91	58.38

Selected:3

M
L

> **Note:** If you cannot see the map underneath, move the attribute table window out of the way.

Q30 Which three states have the lowest percent of households with access to the Internet? You might have to scroll back to the left to see the names of the states.

   1. Mississippi

   2. Arkansas

   3. Louisiana

Q31 In which region of the country are these states located?

   a. North

   b. South

   c. East

   d. West

8. Scroll to the bottom of the table.

9. Click the **last state** (row) in the table (don't use the CTRL key).

10. Hold down the **CTRL** key on your keyboard and click the **two states** above it. The last three states should be highlighted in blue. They will also be highlighted on your map.

Q32 Which three states have the highest percent of households with access to the Internet?

   1. Alaska

   2. New Hampshire

   3. Colorado

11. Close the **Attribute Table**.

 12. Click **Clear All Selections**.

183

Q33 Overall, which part of the country appears to have the most access to the Internet?

    a. North

    b. South

    c. East

    d. West

Once again, let's look at where your state fits in.

 13. Click the **Identify** tool.

14. Find your state on the map.

15. Click your state. The **Identify Results** window appears.

16. Scroll down until you see the **INTACC_PCT** field.

Q34 What is the percent of Internet access for households in your state? Write this number on your worksheet.

17. Close the **Identify Results** window.

Step 7: Save your work and exit AEJEE.

1. Ask your teacher how to save your work.

2. Click the **File** menu and choose **Exit**.

Conclusion

The U.S. Census collects a lot of information about the population. The 2010 Census will have some new collection methods, including handheld computers for enumerators (the people who go door-to-door to collect census data). Questions change with each census as the information required by the U.S. government changes. The most important thing that you can do to help the U.S. Census is to make sure that you get counted!

Using GIS to create maps of the census information can show people how things are spread out across the country.

Module 3, Lesson 2

Patterns of a growing population

Worksheet

Name of your state: Answers will vary.

Topic	Your state
Population in 2000	Answers will vary.
Population density	Answers will vary.
Born in state	Answers will vary.
Born in another state	Answers will vary.
Internet access	Answers will vary.

Once you have completed this table, write a brief paragraph that describes the population characteristics of your state. Use the information you have obtained during this activity.

Answers will vary.

MODULE 4

Analyzing spatial information: U.S. tornadoes

Introduction

While tornadoes occur all over the world, they occur most often in the United States in an area called Tornado Alley. This is a region where people expect tornadoes to strike, but many historical tornado outbreaks also have occurred outside of Tornado Alley. This module puts students in the role of tornado analyst. Using GIS as a tool, students will learn where tornadoes occur most often, where they are the strongest, and where they are the most concentrated. Students will compare tornadoes by their strength and by the seasons of the year. Finally, they will look at the characteristics of some historical tornado outbreaks in the last hundred years.

Lesson 1: Finding Tornado Alley

Lesson 2: Analyzing historical tornadoes

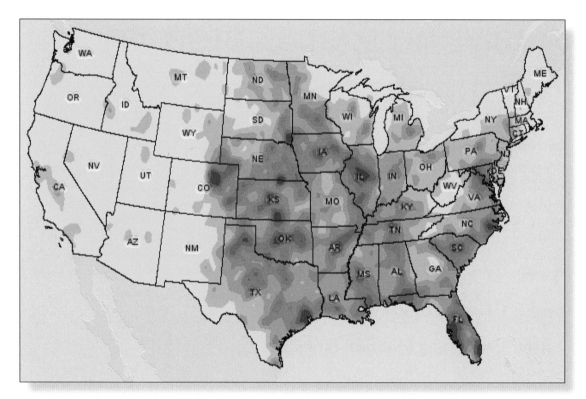

A tornado density map helps students define the area known as Tornado Alley.

Module 4: Lesson 1

Finding Tornado Alley

Lesson overview

Students will work with 10 years of tornado data for the entire United States to locate Tornado Alley. Furthermore, they will use three different approaches—frequency, intensity, and density—to identify Tornado Alley, the part of the country where most tornadoes occur. Finally, students will examine and compare seasonal differences in tornado occurrences.

Estimated time

Approximately 60 minutes

Materials

The student activity and worksheets can be found in the student workbook or on the Software and Data CD. Install the teacher resources on your computer to access them.

Location: OurWorld_teacher\Module4\Lesson1
- Student activity and worksheets: M4L1_student.pdf

Objectives

After completing this lesson, students will be able to do the following:

- Identify and compare regions in the United States
- Compare different map layers and identify relationships between them
- Identify spatial patterns and give reasons to explain them
- Identify temporal patterns and give reasons to explain them
- Analyze and compare the frequency, intensity, and density of tornadoes and learn about the differences between them
- Compare the geographic extent of different layers

GIS tools and functions

Open a map file

Zoom to the geographic extent of a layer that is active (highlighted)

Zoom to the previous map extent

Add a data layer to a map

Clear all selections for all layers in the map

Use the Identify tool to get information about a feature

Use a query to select features and records in a layer

Draw a buffer of a certain distance around selected features

Open the attribute table for an active layer

- Turn layers on and off
- Activate a layer
- Change a symbol
- Sort fields in a table
- Select records in a table
- Create a graduated symbols legend
- Create a unique symbols legend

National geography standard

Standard	K-4	5-8
5 That people create regions to interpret Earth's complexity	The concept of a region as an area of the Earth's surface with unifying geographic characteristics	The influences and effects of regional labels and images
7 The physical processes that shape the patterns of Earth's surface	How patterns (location, distribution, and association) of features on Earth's surface are shaped by physical processes	How to predict the consequences of physical processes on Earth's surface
15 How physical systems affect human systems	How variations within the physical environment produce spatial patterns that affect human adaptation	How natural hazards affect human activities

M ● ● ● ●
L ● ● ○

Teaching the lesson

Introducing the lesson

Begin this lesson by reviewing these concepts:

* How the United States is made up of landforms and physiographic regions
* Three different aspects of tornado occurrence: frequency, intensity, and density
* Averages and how to calculate them
* How studying natural disasters can help you learn about weather, climate, landforms, and other aspects of geography
* How patterns of features on maps have spatial and temporal patterns
* Tornado dynamics
* How weather patterns change by season and by year

Student activity

We recommend that you complete this lesson yourself before completing it with students. This will allow you to modify the activity to accommodate the specific needs of your students.

Teacher notes

* For younger grades, you can conduct the GIS activity as a teacher-led activity with students following along. You can lead students through the GIS steps and ask them the associated questions as a class.
* For older grades, ideally, students will have access to their own computers, but students can complete the activities in groups or under the direction of a teacher.
* Throughout the GIS activity, students are presented with questions. The GIS activity is designed so that students can mark their answers directly on these sheets. Alternatively, you can create a separate answer sheet.
* We recommend that students save their work as they progress through the GIS activity. Students can use either the Save command (to save their changes to the original map) or the Save As command (to save their changes to a new map). Please explain to students where and how they should save their work.

The following are things to look for while students are working on this lesson:

* As students work through the steps, are they thinking about the underlying geographic concepts (e.g., Are they looking for spatial patterns? Are they looking for relationships between features and layers? Are they relating temporal patterns to what they know about seasons of the year?)?
* Are students answering the questions in the GIS activity as they work through the steps?
* Are students able to use the legends to interpret the map layers?
* Do students understand the concept of selecting and sorting records in a table?
* Do students understand the difference between attribute query (selecting features based on an attribute value) and spatial query (selecting features based on location)?

Concluding the lesson

- Engage students in a discussion about the observations and discoveries they made during their exploration of U.S. tornadoes.
- Ask students about their impressions of Tornado Alley.
- Ask students to characterize regions of the country based on what they have learned about U.S. tornadoes.
- Make three copies of the worksheet 2 called "Map of the 48 states." After students complete the activity, ask them to create three maps of Tornado Alley based on frequency, intensity, and density. Have them compare these maps and discuss the differences.
- Has this activity raised any questions that students would like to explore further?
- How can GIS help students to learn about natural disasters?
- Has this activity changed students' ideas about tornadoes?

Extending the lesson

- Ask students to determine how many tornadoes occurred in their state from 1995 to 2004.
- Ask students to determine the date and intensity of the tornado closest to where they live.
- Ask students to determine the number of tornadoes in each state. The following is a way to do this:
  - Select a state
  - Use the Buffer tool to select the tornadoes in that state

References

- http://www.tornadochaser.net/tornalley.html
- http://www.stemnet.nf.ca/CITE/tornadoes_alley.htm
- http://www.ncdc.noaa.gov/oa/climate/severeweather/tornadoes.html
- http://www.aon.com/us/busi/risk_management/risk_consulting/impact_forecast_files/UnitedStatesTornadoSeasonality.pdf
- http://www.photolib.noaa.gov/nssl/tornado1.html

Student activity answer key

Answers appear in blue.

Module 4, Lesson 1

Finding Tornado Alley

Tornado Alley is an area of the United States that has more tornadoes than any other place on earth. Many people argue about which states are in Tornado Alley. In this lesson, you will learn how to find this area for yourself! By the time you finish this lesson, you will be able to list the states that have the most frequent tornadoes, the strongest tornadoes, and the greatest concentration of tornadoes. From your list, you will be able to identify states that are in Tornado Alley.

Tornadoes are associated with certain weather patterns, and these patterns change with the seasons. In this lesson, you will learn which regions of the United States have tornadoes at different times of the year—winter, spring, summer, and fall.

Let's explore Tornado Alley.

Step 1: Start AEJEE.

1. Ask your teacher how to start the AEJEE software.

2. Click the **Maximize** button at the top of the AEJEE window. Now the AEJEE window fills your screen.

Step 2: Open the project.

1. Click the **Open** button.

2. Navigate to your **OurWorld1\Module4** folder.

3. Choose **USTornadoes.axl**.

4. Click **Open**.

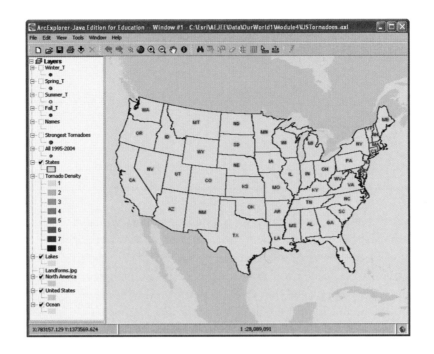

A map of the 48 states labeled with their abbreviations appears on your screen. On the left side of the map, you see a list of layers. Layers are used to show geographic data on a GIS map. Each layer has a name and a legend. You can turn layers on and off. At the top of the map, you see menus and buttons that you will use during this activity.

Step 3: Tornadoes of the United States.

1. Turn on the **All 1995-2004** layer by clicking the box next to the layer name. Each dot represents a tornado that occurred from 1995–2004 (a period of 10 years).

Q1 Look at your map. In what part of the country do most tornadoes occur? (Circle the correct answer.)

    a. Eastern states

    b. Western states

    c. Central and eastern states

Q2 Do you see any states that don't have tornadoes? (Circle the correct answer.)

    a. Yes

    b. No

2. Click the **States** layer name to highlight it. It becomes active.

 3. Click the **Zoom to Active Layer** button to zoom out to all the states. Now you can see Alaska and Hawaii too.

> (Q3) **Do you see any states that don't have tornadoes? (Circle the correct answer.)**
> a. **Yes**
> b. No

 4. Click the **Previous Extent** button to zoom back in to the 48 states.

Tornadoes are more common in some parts of the country than others. Let's look at a different view of the United States.

5. Turn off the **States** layer by clicking the check mark next to the layer name.

6. Turn off the **All 1995-2004** layer.

7. Turn on **Landforms.jpg** by clicking the box next to the layer name. Be patient while this layer draws.

You see an image of the United States showing high areas (mountains) and low areas (plains and valleys).

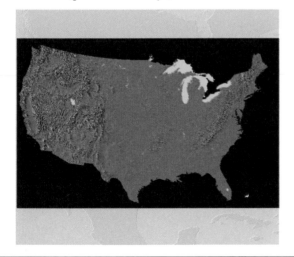

195

8. Turn on the **Names** layer by clicking the box next to the layer name.

Now you see the names of some of the landforms, water bodies, and neighboring countries of the United States.

9. Turn on the **All 1995-2004** layer again.

Q4 Look at your map. In what part of the country do most tornadoes occur? (Circle the correct answer.)

a. **Rocky Mountains**

b. Great Plains

c. **Appalachian Mountains**

The Great Plains have lots of thunderstorms, creating the kind of weather conditions that can cause tornadoes. Thunderstorms occur when warm, moist air meets cool, dry air.

10. Turn off the **Landforms.jpg** layer.

Landforms.jpg

11. Turn on the **States** layer.

Warm, moist air that feeds these thunderstorms comes from the body of water south of the Great Plains.

Q5 Look at your map. What is the name of this body of water? Gulf of Mexico

Cool, dry air comes from the land to the north of the Great Plains.

Q6 Look at your map. What is the name of this country? Canada

12. Turn off the **Names** layer.

13. Turn off the **All 1995-2004** layer.

M ● ● ● ●
L ● ● ○

Step 4: States where tornadoes occur most often.

While Tornado Alley is generally in the middle of the country, many people argue about the specific states that belong in this area. There are three different ways, or methods, to decide which states are in Tornado Alley:

1. States where tornadoes occur most often
2. States where the most dangerous tornadoes occur
3. States where the greatest concentration of tornadoes occurs

You will use GIS to analyze these three methods and decide on some of the states that are in Tornado Alley. You will start by finding out where tornadoes occur most often.

1. Click the **Add Data** button.

2. Navigate to your **OurWorld1\Module4\Data** folder.

3. Click **State_tornadoes.shp**.

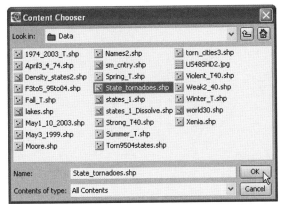

4. Click **OK**. The **State_tornadoes** layer is added to the map and draws in a random color.

   The State_tornadoes layer contains information about the tornadoes for each state. You will change its symbols.

5. Click the **State_tornadoes** layer name to highlight it. It becomes active.

6. Right-click the **State_tornadoes** layer name and choose **Properties**. This brings up the **Properties** window.

197

7. Under **Draw features using**, choose **Graduated Symbols**.

8. For **Field**, choose **AVERAGE_T**. This field contains the average number of tornadoes per year in each state.

9. Click **OK**.

You see a map showing the average number of tornadoes per year in each state (from 1995 to 2004). The red color means there were a lot of tornadoes; the yellow means there were relatively few tornadoes.

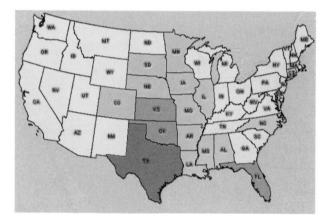

What does "average number of tornadoes per year" mean?

Suppose a state had 200 tornadoes from 1995 to 2004 (a period of 10 years). If you divide the total number of tornadoes (200) by the number of years (10), you will get the average number of tornadoes per year:

$200 \div 10 = 20$ tornadoes per year

Q7 **Look at your map and at the legend. In which state did tornadoes occur most often?** Texas

M ● ● ● ●
L ● ● ○

10. Right-click **State_tornadoes** and choose **Attribute Table**. This brings up the Attribute Table.

11. Scroll to the right in the table until you see the **AVERAGE_T** field.

12. Right-click the **AVERAGE_T** field name and choose **Sort Descending**. The states are listed in order from the highest number to the lowest number of tornadoes.

Attributes of State_tornadoes					
_FIPS	STATE_ABBR	SQMI	TOTAL_T	AVERAGE_T	STRO_VIOL
	HI	6381	4	0	0
	WA	67290	33	3	Sort Ascending
	MT	147245	107	1	Sort Descending
	ME	32162	14	1	Sort Selected Data to Top
	ND	70812	324	3	Export Selected Data
	SD	77195	332	33	8
	WY	97803	116	12	0
	WI	56088	184	18	7

Let's look at the top six states.

13. Hold down your **CTRL** key on the keyboard, and slowly click each of the first **six** records in the table (count them). They become highlighted in blue. They also become highlighted on your map.

Attributes of State_tornadoes					
_FIPS	STATE_ABBR	SQMI	TOTAL_T	AVERAGE_T	STRO_VIOL
	TX	264436	1636	164	31
	KS	82197	811	81	24
	FL	55815	783	78	4
	OK	70003	667	67	24
	IL	56299	643	64	18
	NE	77330	625	62	8
	IA	56258	550	55	16
	CO	104101	516	52	4
	MN	84520	450	45	8
	AR	52913	454	45	35
	AL	51716	405	40	14
	MO	69833	385	38	17
	MS	47619	384	38	20

Selected:6

> **Note:** If you make a mistake, click the first record without holding down the CTRL key. Then hold down the CTRL key and click the next five records.

14. Let go of the **CTRL** key.

15. Scroll to the left in the table until you see the **STATE_NAME** field. You see the names of six states highlighted in blue.

16. Take out the "Tornado Alley" worksheet that your teacher gave you.

17. Find Column 1: Highest average number of tornadoes.

Q8 Write the names of these six states in Column 1 of your worksheet. (Some state names are filled in for you.)

18. Close the **Attribute Table**.

 19. Click the **Clear All Selections** button.

Step 5: States where the strongest tornadoes happen.

Another way to decide which states are in Tornado Alley is to see where the strongest tornadoes happen. The strength of a tornado is measured by the speed of its winds and by how much damage it causes. You will look at which states had the strongest tornadoes between 1995 and 2004.

1. Turn off the **State_tornadoes** layer. The layer name should still be highlighted.

2. Turn on the **Strongest Tornadoes** layer.

Each purple dot represents a single tornado. These tornadoes had winds that were stronger than 150 miles per hour and caused a lot of damage. Only a small portion of all tornadoes are really strong.

Q9 Look at the map. List two states that you think have a high number of strong tornadoes. Answers may include Arkansas, Tennessee, Oklahoma, Texas, or Kansas.

> **Note:** If you aren't sure of the name of a state, click the Identify tool 🛈 , then click the state. Close the Identify Results window when you are finished.

Now you will find out if you were right.

3.   Turn off **Strongest Tornadoes**.

4.   Turn on **State_tornadoes**. The layer name should still be highlighted.

5.   Right-click **State_tornadoes** and choose **Properties**. This brings up the **Properties** window.

6.   Change the **Field** to **STRO_VIOL**. This field contains the strongest tornadoes from 1995 to 2004.

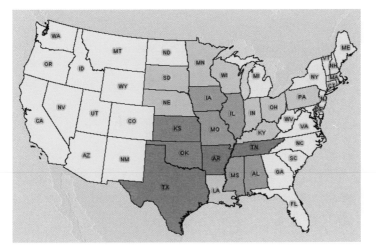

7.   Click **OK**.

The red color means more strong and violent tornadoes, the yellow color means less.

 8.   Click the **Identify** tool.

9.   Click one of the **dark red states**. The **Identify Results** window opens. On the left side of the window you see the name of the state. On the right side you see field names and values.

10. Look at the value for the **STRO_VIOL** field. It tells you the number of strong and violent tornadoes that occurred in that state from 1995 to 2004.

 What is the name of this state? Answers may include Texas, Arkansas, or Tennessee.

 How many strong and violent tornadoes did it have from 1995 to 2004? Answers may include 31 (Texas), 35 (Arkansas), or 30 (Tennessee).

11. Close the **Identify Results** window.

12. Click another **dark red state** on the map.

 What is the name of this state? Answers may include Texas, Arkansas, or Tennessee.

 How many strong and violent tornadoes did it have from 1995 to 2004? Answers may include 31 (Texas), 35 (Arkansas), or 30 (Tennessee).

13. Close the **Identify Results** window.

14. Right-click the **State_tornadoes** layer name and choose **Attribute Table**.

15. Scroll to the right side of the table.

16. Right-click the **STRO_VIOL** field name and choose **Sort Descending**.

STATE_ABBR	SQMI	TOTAL_T	AVERAGE_T	STRO_VIOL
HI	6381	4	0	0
WA	67290	33	3	0
MT	147245	107	11	0
ME	32162	14	1	0
ND	70812	324	32	4
SD	77195	332	33	8
WY	97803	116	12	0

The states are listed in order from the highest number of strong and violent tornadoes to the lowest number.

Let's look at the top six states.

17. Hold down your **CTRL** key on your keyboard and click the first **six** records in the table (count them). They become highlighted in blue. They also become highlighted on your map.

Note: If you make a mistake, click the first record without holding down the CTRL key. Then hold down the CTRL key and click the next five records.

18. Let go of the **CTRL** key.

# Attributes of State_tornadoes					
PS	STATE_ABBR	SQMI	TOTAL_T	AVERAGE_T	STRO_VIOL
	AR	52913	454	45	35
	TX	264436	1636	164	31
	TN	42092	278	28	30
	KS	82197	811	81	24
	OK	70003	667	67	24
	MS	47619	384	38	20
	IL	56299	643	64	18
	MO	69833	385	38	17
	IA	56258	550	55	16
	AL	51716	405	40	14
	KY	40320	208	21	13

Selected:6

19. Scroll to the left in the table until you see **STATE_NAME**. You see the names of six states highlighted in blue.

20. Take out your "Tornado Alley" worksheet.

21. Find Column 2: Highest number of strong or violent tornadoes.

Q14 Write the names of these six states in Column 2 of your worksheet. (Some state names are filled in for you.)

22. Close the **Attribute Table**.

 23. Click the **Clear All Selections** button.

Step 6: States with the highest concentration of tornadoes.

Looking at tornado density is a third way to decide which states are in Tornado Alley. This shows where the most tornadoes occured within the smallest area of land (closest together). When you map how close together or far apart tornadoes are, you call this a tornado "density" map.

1. Turn off **State_tornadoes**.

2. Turn on the **Tornado Density** layer.

The states cover up the Tornado Density layer, so you will change the symbol for the states.

203

3. Right-click the **States** layer name and choose **Properties**. This brings up the **Properties** window.

4. Change the **Style** to **Transparent fill**.

5. Click **OK**.

Now you can see the state outlines and labels on top of the Tornado Density layer.

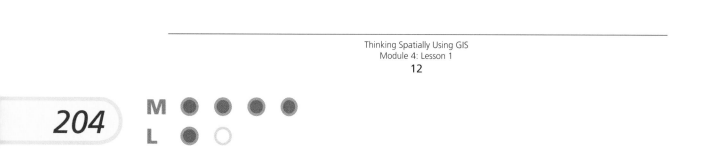

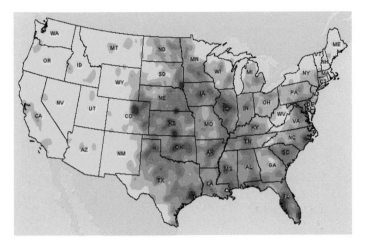

The dark red areas show where tornadoes are most dense (close together). The light red (pink) areas show where tornadoes are least dense (spaced far apart).

6. Click the **Tornado Density** layer name to highlight it. It becomes active.

7. Click the **Query Builder** button.

8. Under **Select a field**, click **DENSITY**.

9. Click the > (greater than) button.

10. Under **Values**, click **6**.

11. Click **Execute**.

12. Close the **Query Builder** window.

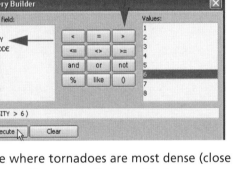

13. Look at the map. The highlighted areas are where tornadoes are most dense (close together). You will use the GIS to draw a one-mile area (called a buffer) around each yellow area. Then you will use these buffers to select the states in the layer beneath them.

> **Buffer** is a common GIS operation that draws an area of a certain distance around a feature. For instance, you might draw a buffer of a certain distance around a school to show a safety zone.

14. Click the **Buffer** button.

15. Click the box next to **Use buffer to select features from this layer**.

A list appears below the box.

16. Choose **States** from the list.

17. Click **OK**.

The states containing dense tornado areas are selected on the map.

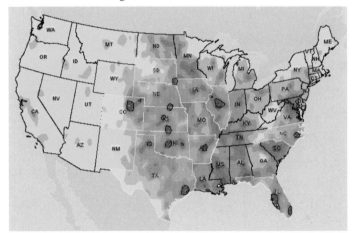

18. Click the **States** layer name to highlight it. It becomes active.

19. Click the **Attributes** button. The **Attributes of States** table opens.

20. Right-click **STATE_NAME**, and choose **Sort Selected Data to Top**.

You see selected states highlighted in blue at the top of the table.

21. Look at the bottom of the table where it says **Selected**.

Q15 **How many states contain dense tornado areas?** 11

22. Take out your "Tornado Alley" worksheet.

23. Find Column 3: Highest tornado density.

Q16 **Write the names of the first six states in Column 3 of your "Tornado Alley" worksheet. (The rest of the state names are already filled in for you.)**

24. Close the **Attribute Table**.

 25. Click the **Clear All Selections** button.

26. Turn off the **Tornado Density** layer.

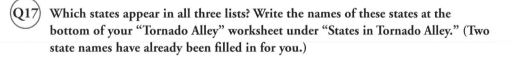

You have defined Tornado Alley in three different ways:
1. States where tornadoes occur most often
2. States where the strongest tornadoes occur
3. States where tornado density is highest

You listed these states on your "Tornado Alley" worksheet in Columns 1, 2, and 3.

27. Look at Columns 1, 2, and 3 now.

Q17 **Which states appear in all three lists? Write the names of these states at the bottom of your "Tornado Alley" worksheet under "States in Tornado Alley." (Two state names have already been filled in for you.)**

28. Take out your "Map of the 48 states" worksheet.

29. Color in these six states on the map.

Over 10 years, from 1995 to 2004, these states have been in Tornado Alley no matter what method you use to define this area.

207

Step 7:   Where tornadoes occur during the seasons.

You have seen how tornadoes can occur close together over land. They can also occur close together over time. In this step you will look at where tornadoes occur during each season of the year and in each region of the United States.

1. Right-click the **States** layer name and choose **Properties**.

2. Under **Draw features using**, choose **Unique Symbols**.

3. For **Field for values**, scroll to the end of the list and choose **Region** (not SUB_REGION).

4. For **Color Scheme**, choose **Minerals**.

5. Click **OK**.

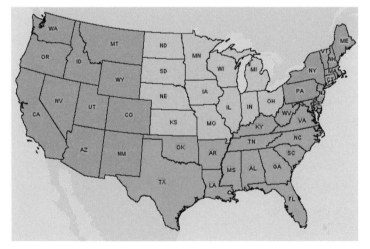

You see four United States regions, each with a different color, and you can see the state outlines.

6. Turn on the **Winter_T** layer.

Winter tornadoes from 1995 to 2004 appear as green dots on the map.

Q18 Look at the map. In what region of the country are most winter tornadoes? Hint: Look at the States legend to see the names and colors of the regions. (Circle the correct answer.)

    a. **Midwest**

    b. **Northeast**

    c. South

    d. **West**

7. Click the **Winter_T** layer name to highlight it. It will become active.

8. Right-click the **Winter_T** layer name and choose **Attribute Table**.

The FID field contains a number for each winter tornado.

9. Scroll to the bottom of the table to see the highest FID number.

Q19 How many tornadoes occurred in winter between 1995 and 2004? 899

10. Close the **Attribute Table**.

11. Turn off the **Winter_T** layer.

12. Turn on the **Spring_T** layer.

Spring tornadoes from 1995 to 2004 appear as red dots on the map.

Q20 In what regions of the country are most spring tornadoes? (Circle the correct answer.)

    a. **Midwest**

    b. **Northeast**

    c. South

    d. **West**

13. Click the **Spring_T** layer to make it active.

14. Right-click the **Spring_T** layer name and choose **Attribute Table**.

Module 4: Lesson 1

209

15. Scroll to the bottom of the table to see the highest FID number.

 How many tornadoes occurred in spring between 1995 and 2004? 5,285

16. Close the **Attribute Table**.

 Look at the map. List two differences between winter tornadoes and spring tornadoes. (Hint: You can turn both layers on at the same time or turn them on and off to compare them.) Answers may include the following:

- There are many more tornadoes in the spring.
- Winter tornadoes occur mostly in the South, with very few in the Midwest or Northeast.
- Spring tornadoes extend much farther north and west.
- Spring tornadoes cover a much bigger area of the United States.

17. Turn off the **Spring_T** layer.

18. Turn on the **Summer_T** layer.

Summer tornadoes from 1995 to 2004 appear as yellow dots on the map.

 In what region of the country are most summer tornadoes? (Circle the correct answer.)

  a. Midwest

  b. Northeast

  c. South

  d. West

 Look at the map. List one difference between summer tornadoes and spring tornadoes. (Hint: You can turn both layers on at the same time or turn them on and off to compare them.) Answers may include the following:

- Summer tornadoes are more concentrated in the northern states.
- Spring tornadoes are more concentrated in the southern states.

19. Click the **Summer_T** layer to make it active.

☐ ☑ Summer_T
   └ ○

210

M ● ● ● ●
L ● ● ○

20. Right-click the **Summer_T** layer name and choose **Attribute Table.**

21. Scroll to the bottom of the table to see the highest FID number.

 Q25 **How many tornadoes occurred in summer between 1995 and 2004?** 4,383

22. Close the **Attribute Table**.

23. Turn off the **Summer_T** layer.

24. Turn on the **Fall_T** layer.

Fall tornadoes from 1995 to 2004 appear as blue dots on the map.

 Q26 **In what region of the country are most fall tornadoes? (Circle the correct answer.)**
   a. **Midwest**
   b. **Northeast**
   c. South
   d. **West**

25. Click the **Fall_T** layer name to make it active.

26. Right-click the **Fall_T** layer name and choose **Attribute Table**.

27. Scroll to the bottom of the table to see the highest FID number.

 Q27 **How many tornadoes occurred in fall between 1995 and 2004?** 2,173

28. Close the **Attribute Table**.

Q28 **Which season had the most tornadoes?**
   a. **Winter**
   b. Spring
   c. **Summer**
   d. **Fall**

211

Module 4: Lesson 1

(Q29) What is the peak season for the southern states (the season when the South has the most tornadoes)?

a. Winter

b. Spring

c. Summer

d. Fall

(Q30) What is the peak season for the northern states (the season when they have the most tornadoes)?

a. Winter

b. Spring

c. Summer

d. Fall

(Q31) Discuss with your classmates some possible reasons that the locations of tornadoes change with the season. Write down two possible reasons. Answers may include the following:

• The temperature warms in the spring and summer.

• More thunderstorms occur in the spring and summer.

• The sun is farther north in the sky during spring and summer.

• The temperature difference between warm and cold air masses is greater during the spring and summer, producing more thunderstorms.

Step 8: Save your work and exit AEJEE.

1. Ask your teacher where and how to save your work.

2. Click the **File** menu and choose **Exit**.

M

L

Conclusion

People who live in Tornado Alley are aware of the dangers of tornadoes and understand that they are part of life.

GIS can help researchers and scientists map patterns that show where tornadoes touch down and study how tornadoes behave on the ground, including how far and fast they move and how wide an area they cover. Their research leads to better warning systems and better methods for predicting when a tornado might occur. These improvements make it safer for people who live in Tornado Alley.

Module 4, Lesson 1

Finding Tornado Alley

Worksheet 1: Tornado Alley

Column 1	Column 2	Column 3
Highest average number of tornadoes	Highest number of strong or violent tornadoes	Highest tornado density
1. Texas	1. Arkansas	1. South Dakota
2. Kansas	2. Texas	2. Iowa
3. Florida	3. Tennessee	3. Nebraska
4. Oklahoma	4. Kansas	4. Illinois
5. Illinois	5. Oklahoma	5. Colorado
6. Nebraska	6. Mississippi	6. Kansas
7. Iowa	7. Illinois	7. Oklahoma
8. Colorado	8. Missouri	8. North Carolina
9. Minnesota	9. Iowa	9. Texas
10. Arkansas	10. Alabama	10. Arkansas
11. Alabama	11. Kentucky	11. Florida

States in Tornado Alley
1. Texas
2. Kansas
3. Oklahoma
4. Illinois
5. Iowa
6. Arkansas

214

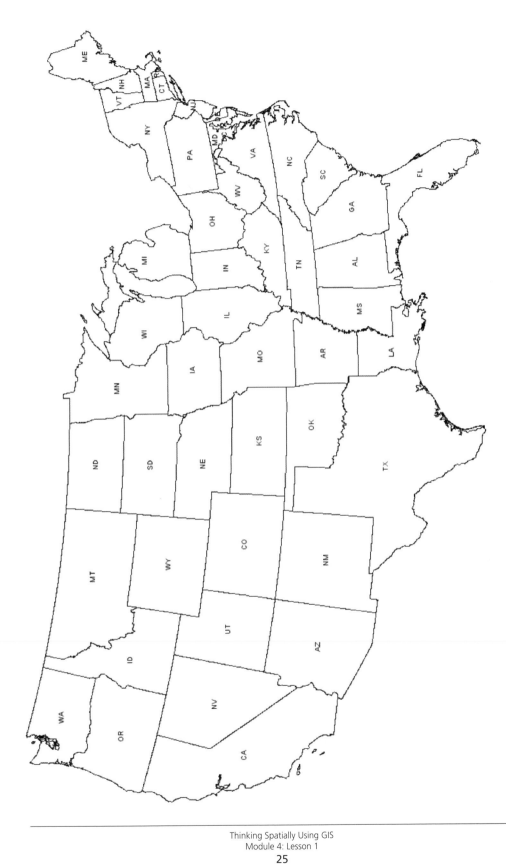

Name: _____   Date: _____

Worksheet 2: Map of the 48 states

215

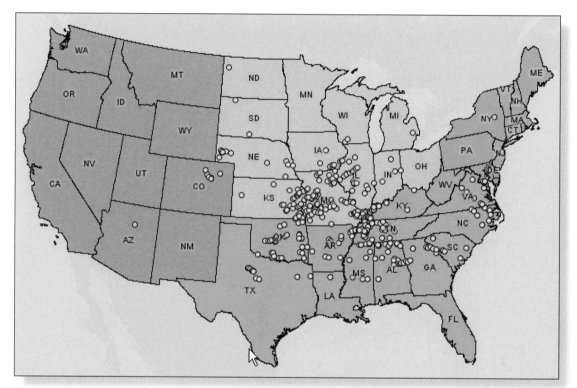

Between May 1 and May 10, 2003, more than 400 tornadoes hit the central and southern regions of the United States.

Module 4: Lesson 2

Analyzing historical tornadoes

Lesson overview

Students will analyze U.S. tornadoes to compare some of the most well-known tornado outbreaks in history. Students will work with more than 30 years of tornado data for the entire United States. In this lesson, students will use the Fujita Tornado Damage Scale to classify tornadoes and identify the amount and types of damage associated with tornadoes of different strengths. Finally, they will look for spatial patterns and compare three well-known tornado outbreaks.

Estimated time

Approximately 60 minutes

Materials

The student activity, handout, and worksheet can be found in the student workbook or on the Software and Data CD. Install the teacher resources on your computer to access them.

Location: OurWorld_teacher\Module4\Lesson2
* Student activity, handout, and worksheet: M4L2_student.pdf

Additional materials
* Calculator

Objectives

After completing this lesson, students will be able to do the following:

* Identify and compare regions in the United States
* Compare different map layers and identify relationships between them
* Identify spatial patterns and give reasons to explain them
* Identify temporal patterns and give reasons to explain them
* Analyze and compare the frequency and intensity of tornadoes and learn the differences between them
* Compare the geographic extent of different layers

GIS tools and functions

- Open a map file
- Zoom to the geographic extent of a layer that is active (highlighted)
- Zoom to the previous map extent
- Use a query to select features and records in a layer
- Clear all selections for all layers in the map
- Draw a buffer of a certain distance around selected features
- Open the attribute table for an active layer
- Select features on the map using the mouse
- Turn layers on and off
- Activate a layer
- Change a symbol
- Sort fields in a table
- Select records in a table
- Create a graduated color legend
- Get statistics for selected records

National geography standards

Standard	K-4	5-8
5 That people create regions to interpret Earth's complexity	The concept of a region as an area of the Earth's surface with unifying geographic characteristics	The influences and effects of regional labels and images
7 The physical processes that shape the patterns of Earth's surface	How patterns (location, distribution, and association) of features on Earth's surface are shaped by physical processes	How to predict the consequences of physical processes on Earth's surface
15 How physical systems affect human systems	How variations within the physical environment produce spatial patterns that affect human adaptation	How natural hazards affect human activities

Teaching the lesson

Introducing the lesson

Begin this lesson by reviewing these concepts:

- Natural disasters: where they occur and how they affect people all over the country
- How natural disasters can cause a significant amount of damage
- How studying natural disasters can help you learn about weather, climate, landforms, and other aspects of geography
- Tornado dynamics
- Tetsuya Fujita and the Storm Prediction Center

Student activity

We recommend that you complete this lesson yourself before completing it with students. This will allow you to modify the activity to accommodate the specific needs of your students.

Teacher notes

- For younger grades, you can conduct the GIS activity as a teacher-led activity in which students follow along. You can lead students through the GIS steps and ask them the associated questions as a class.
- For older grades, ideally each student will have access to a computer, but students can complete the activities in groups or under the direction of a teacher.
- Throughout the GIS activity, students are presented with questions. The GIS activity is designed so that students can mark their answers directly on these sheets. Alternatively, you can create a separate answer sheet.
- We recommend that students save their work as they progress through the GIS activity. Students can use either the Save command (to save their changes to the original map) or the Save As command (to save their changes to a new map). Please explain to students where and how they should save their work.
- Throughout this activity, students are asked to calculate percentages. Students can use either a handheld calculator or the calculator on their computer.
- The tornado data used in this lesson was downloaded from the National Atlas Web site (http://www.nationalatlas.com/mld/tornadx.html). The actual number of tornadoes and deaths from tornadoes recorded in this dataset may differ slightly from reports found on the Internet. For example, it is generally reported that 401–412 tornadoes occurred from May 1–10, 2003. In the tornado data used in this lesson, the number of tornadoes recorded for the May 1–10, 2003, event is 366.

The following are things to look for while students are working on this lesson:

- As students work through the steps, are they thinking about the underlying geographic concepts (e.g., are they looking for spatial patterns? Are they looking for relationships between features and layers? Are they relating the map to the information on their worksheet?)
- Are students answering the questions in the GIS activity as they work through the steps?
- Are students able to use the legends to interpret the map layers?
- Do students understand the concept of selecting features from a layer?
- Do students understand the difference between attribute query (selecting features based on an attribute value) and spatial query (selecting features based on location)?

Concluding the lesson

- Engage students in a discussion about the observations and discoveries they made during their exploration of 30 years of U.S. tornadoes and noted tornado outbreaks.
- Ask students to characterize regions of the country based on what they have learned about U.S. tornadoes.
- Has this activity raised any questions that students would like to explore further?
- Ask students to research the terms "tornado watch" and "tornado warning."
- Ask students if their community or school has a tornado plan and what it contains.
- How can GIS help students to learn about natural disasters?
- Has this activity changed students' ideas about tornadoes?

Extending the lesson

- Have students create a graph comparing the characteristics of the three tornado outbreaks covered in this lesson, such as total number of tornadoes, number of states affected, and total number of strong and violent tornadoes.
- Ask students to determine the date and intensity of the tornado closest to where they live.
- Ask students to examine unexpected tornadoes, such as tornadoes occurring in the mountains of Colorado or the canyon lands of Utah.

References

- http://www.spc.noaa.gov/faq/tornado/f-scale.html
- http://www.noaanews.noaa.gov/stories/s345.htm
- http://www.nssl.noaa.gov/edu/safety/tornadoguide.html
- http://www.aon.com/us/busi/risk_management/risk_consulting/impact_forecast_files/UnitedStatesTornadoSeasonality.pdf

Student activity answer key

Answers appear in blue.

Module 4, Lesson 2

Analyzing historical tornadoes

Meteorologists study storms that produce tornadoes. They use tools such as forecasting (predicting the weather), research (observing and trying to understand the weather), and storm chasing (following and watching storms close-up). Some tornadoes and tornado outbreaks may stick even in your memory. Some students grow up to study the science of tornadoes because of a personal experience or out of curiosity about a particular outbreak. Perhaps you will decide to be a meteorologist yourself one day!

In this GIS activity, you will see how tornadoes are classified into weak, strong, and violent categories. You will also take a close look at some of the memorable tornado outbreaks in history.

Step 1:   Start AEJEE.

1. Ask your teacher how to start the AEJEE software.

2. Click the **Maximize** button at the top of the AEJEE window. AEJEE will now take up all the space on your screen.

Step 2:   Open the project.

1. Click the **Open** button.

2. Navigate to your **OurWorld1\Module4** folder.

3. Choose **Historical.axl**.

4. Click **Open**.

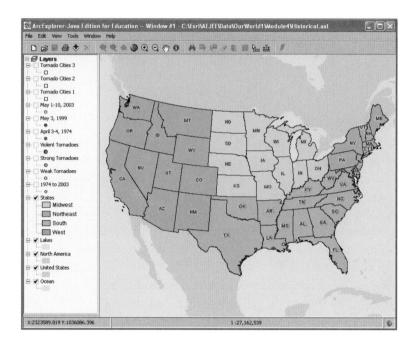

You see states in four regions of the United States. The States legend shows you the name of each region.

Step 3: Thirty years of tornadoes.

From 1974 to 2003, the United States had three memorable tornado outbreaks. A tornado outbreak occurs when many tornadoes form from a single storm. Before looking at these particular outbreaks, you will look at the tornadoes in this 30-year time period.

1. Turn on the **1974 to 2003** layer by clicking the box next to the layer name.

Q1   Which region had the fewest tornadoes between 1974 and 2003?
a   Midwest
b. Northeast
c. South
d. West

2. Click the **1974 to 2003** layer name to highlight it. It becomes active.

 3. Click the **Zoom to Active Layer** button.

Now you see all the states, including Alaska and Hawaii.

Q2 Which state had no tornadoes between 1974 and 2003? Alaska

 4. Click the **Previous Extent** button to zoom back to the 48 states.

5. Right-click the **1974 to 2003** layer name and choose **Attribute Table**. (Be patient while the table opens.)

The first field in the table, called FID (which stands for feature identification number), contains a unique number for every tornado.

FID	#SHAPE#	TORNADX020	YE...	NUM
1	Point	14985	1974	1
2	Point	14986	1974	2
3	Point	14987	1974	3
4	Point	14988	1974	4
5	Point	14989	1974	5
6	Point	14990	1974	6
7	Point	14991	1974	7
8	Point	14992	1974	8
9	Point	14993	1974	9
10	Point	14994	1974	10
11	Point	14995	1974	11
12	Point	14996	1974	12

Selected:0

Click on the scroll bar, *hold* down the mouse button, and drag the bar to the bottom.

6. Scroll to the bottom of the table.

7. Look at the number in the FID field.

Q3 How many tornadoes occurred between 1974 and 2003? (Write the number here. You will use this number in the next step.) 29,640

8. Close the **Attribute Table**.

Step 4: Weak tornadoes.

A tornado is like a spinning wind tunnel. It extends from the thundercloud in the sky to the ground below. (To be a tornado, it must touch the ground.) The strength of a tornado is measured by the speed of the spinning wind and the amount of damage to things on the ground—houses, cars, trains, trees, and roads. Tornado strength is measured on a scale called the Fujita Tornado Damage Scale, or F-Scale for short.

1. Take out the "Tornado F-scale" handout that your teacher gave you. It shows six different strengths, from the weakest (F0) to the most violent (F5). For each size, you see the wind speed and a description of the damage.

    Tornadoes may also be described as weak, strong, or violent:
    *   F0 and F1 tornadoes are considered weak.
    *   F2 and F3 tornadoes are considered strong.
    *   F4 and F5 tornadoes are considered violent.

2. Return to the map on your computer screen.

3. Turn off the **1974 to 2003** layer by clicking the check mark next to the layer name.

4. Turn on **Weak Tornadoes** by clicking the box next to the layer name.

    You see all the weak (F0 and F1) tornadoes that occurred between 1974 and 2003. Weak tornadoes cause very few deaths. Weak tornadoes can last from about 1 minute to 10 minutes. Their winds are from 40 to 112 miles per hour.

Q4   **Which region had the fewest weak tornadoes between 1974 and 2003?**
  a. **Midwest**

  b. **Northeast**

  c. **South**

  d. West

Q5   **Look at your "Tornado F-scale" handout. Write down two examples of damage that occurs with F0 tornadoes.** Answers may include the following:
  *   Chimney damage

  *   Broken twigs and tree branches

- Trees pushed over

- Damage to signboards

- Broken windows

Q6 **Look at your "Tornado F-scale" handout again. Write down two examples of damage that occurs with F1 tornadoes.** Answers may include the following:

- Roof damage

- Mobile homes pushed off foundations

- Autos pushed off roads

- Broken trees

5. Right-click the **Weak Tornadoes** layer name and choose **Attribute Table**.

The first field in the table is the FID (feature identification) field. It contains a unique number for each tornado. The highest FID number tells you the total number of tornadoes in this layer.

FID	#SHAPE#	TORNADX020	YE...	NUM
1	Point	14990	1974	6
2	Point	14991	1974	7
3	Point	14992	1974	8
4	Point	14993	1974	9
5	Point	14999	1974	15
6	Point	15008	1974	24
7	Point	15009	1974	25
8	Point	15013	1974	29
9	Point	15014	1974	30
10	Point	15015	1974	31
11	Point	15016	1974	32
12	Point	15017	1974	33
13	Point	15018	1974	34

Attributes of Weak Tornadoes

Selected:0

6. Scroll to the bottom of the table.

Q7 **How many weak tornadoes occurred in the United States between 1974 and 2003?** 24,858

Module 4: Lesson 2

225

Q8 What percentage of all tornadoes were weak? To answer this question, you will divide the number of weak tornadoes by the total number of tornadoes (you wrote this number down in Q3). Then you will multiply the answer by 100. You may need to use a calculator.

Weak tornadoes ÷ total tornadoes = 0.83 X 100 = 83 percent (without the decimal)

7. Close the **Attribute Table**.

Step 5: Strong tornadoes.

Strong tornadoes are responsible for nearly one-third of all the deaths caused by tornadoes. Strong tornadoes can last 20 minutes or longer. Their winds are from 113 to 206 miles per hour.

How do the wind speeds of strong tornadoes compare to other speeds?
- A cheetah, the fastest land animal, can run as fast as 70 miles per hour.
- The fastest roller coaster in the world can go 120 miles per hour.
- The fastest road car can reach speeds of 240 miles per hour.
- The fastest NASA jet plane can travel up to 2,274 miles per hour.

1. Turn off the **Weak Tornadoes** layer.

2. Turn on the **Strong Tornadoes** layer.

You see all the strong (F2 and F3) tornadoes that occurred from 1974 to 2003.

Q9 Which region had the fewest strong tornadoes between 1974 and 2003?

a   Midwest

b. Northeast

c. South

d. West

Q10 Look at the map. Does it look like there were more strong tornadoes or weak tornadoes from 1974 to 2003? (Hint: Turn the layers on and off to compare them.)

   a. More strong tornadoes

   b. More weak tornadoes

Q11 Look at your "Tornado F-scale" handout. Write two examples of damage that occurs with F2 tornadoes. Answers may include the following:

- Roofs torn off
- Mobile home destruction
- Houses lifted and moved
- Railroad cars pushed over
- Large trees snapped or pulled out
- Light objects take flight

Q12 Look at your "Tornado F-scale" handout. Write two examples of damage that occurs with F3 tornadoes. Answers may include the following:

- Roofs and walls torn off
- Trains turned over
- Forest trees pulled out
- Heavy cars lifted and thrown
- Pavement blown off

3. Right-click the **Strong Tornadoes** layer name and choose **Attribute Table**.

The FID (feature identification) field contains a unique number for each tornado. The highest number tells you the total number of tornadoes in this layer.

FID	#SHAPE#	TORNADX020	YE...	NUM	
1	Point	14985	1974	1	
2	Point	14986	1974	2	
3	Point	14987	1974	3	
4	Point	14988	1974	4	
5	Point	14989	1974	5	
6	Point	14994	1974	10	
7	Point	14995	1974	11	
8	Point	14996	1974	12	
9	Point	14997	1974	13	
10	Point	14998	1974	14	
11	Point	15000	1974	16	
12	Point	15001	1974	17	

Selected:0

4.   Scroll to the bottom of the table.

Q13 **How many strong tornadoes occurred between 1974 and 2003?** 4,517

Q14 **What percentage of all tornadoes were strong? To answer this question, you will divide the number of strong tornadoes by the total number of tornadoes (you wrote this number down in Q3). Then, you will multiply the answer by 100. You may need to use a calculator.**

Strong tornadoes ÷ total tornadoes = 0.15 X 100 = 15 percent (without the decimal)

5.   Close the **Attribute Table**.

Step 6: Violent tornadoes.

Violent tornadoes are responsible for more than two-thirds of all the deaths caused by tornadoes. A violent tornado can last more than one hour. The winds from a violent tornado range from 207 to 318 miles per hour. The most violent tornadoes create the strongest known winds on earth!

1.   Turn off **Strong Tornadoes**.

Strong Tornadoes

2.   Turn on **Violent Tornadoes**.

Violent Tornadoes

You see all the violent (F4 and F5) tornadoes that occurred between 1974 and 2003.

Q15 What region had the fewest violent tornadoes between 1974 and 2003?

   a. **Midwest**

   b. **Northeast**

   c. **South**

   d. West

Q16 Which regions had the most violent tornadoes between 1974 and 2003? (Circle all the correct answers.)

   a. Midwest

   b. **Northeast**

   c. South

   d **West**

Q17 Were there more violent tornadoes or strong tornadoes from 1974 to 2003? (Hint: Turn the layers on and off to compare them.)

   a. **More violent tornadoes**

   b. More strong tornadoes

Q18 Look at your "Tornado F-scale" handout. Write two examples of damage that occurs with F4 tornadoes. Answers may include the following:

- Flattened houses
- Weak buildings blown away
- Cars thrown
- Large objects take flight
- Forest trees blown away

Q19 Look at your "Tornado F-scale" handout. Write two examples of damage that occurs with F5 tornadoes. Answers may include the following:

- Strong houses blown away
- Very large objects (like cars) take flight
- Bark blown off trees
- Incredible things happen

Module 4: Lesson 2

229

3. Right-click the **Violent Tornadoes** layer name and choose **Attribute Table**.

The FID field contains a unique number for each tornado. The highest number tells you the total number of tornadoes in this layer.

4. Scroll to the bottom of the table.

Q20 How many violent tornadoes occurred between 1974 and 2003? 265

Q21 What percentage of all tornadoes were violent? You already calculated the percentage of weak and strong tornadoes. Write those down here.

Percentage of weak tornadoes: __ __percent (from Q8)

Percentage of strong tornadoes: __ __percent (from Q14)

Add these two numbers together: __ __percent
Subtract this number from 100%.

What is the percentage of violent tornadoes?

a. 1 percent

b. 2 percent

c. 3 percent

d. 4 percent

5. Close the **Attribute Table**.

Step 7: The most violent tornado outbreak that we know about in history happened on April 3–4, 1974.

1. Turn off the **Violent Tornadoes** layer.

2. Turn on the **April 3–4, 1974** layer.

Your map shows all the tornadoes that took place in these two days.

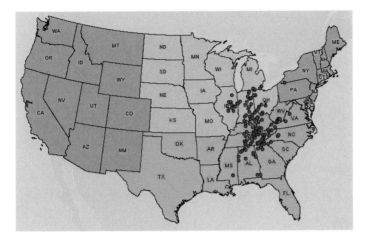

3. Turn on the **Tornado Cities 1** layer. You see the city of Xenia, Ohio, where the worst tornado took place.

4. Right-click the **Tornado Cities 1** layer name and choose **Attribute Table**. You see one record (row) in the table.

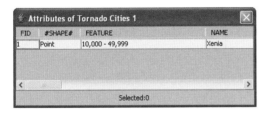

5. Scroll to the right in the table until you see the **POP_2000** field. This field contains the population for the year 2000.

 Q22 How many people lived in Xenia in 2000? 24,164

6. Close the **Attribute Table**.

The tornado that hit Xenia was an F5. It destroyed half the town and caused 100 million dollars in property damage.

7. Take out the "Historical tornadoes" worksheet that your teacher gave you.

8. Look at the column called **EVENT 1**. There is a lot of information about the April 3–4, 1974, tornado outbreak. Use this information to answer the questions below.

 Q23 How many tornadoes occurred during this 24-hour period? 148

231

Q24 How many states were affected? 13

Q25 How many F5 tornadoes occurred on April 3 and 4? 7

The total number of injuries and deaths is not filled in. You will use the GIS to get this information.

9. Turn off the **Tornado Cities 1** layer.

10. Click the **April 3-4, 1974** layer name to make it active. It becomes highlighted.

11. Click the **Query Builder** button.

12. Under **Select a field**, scroll down and click **INJ**.

13. Click the **>** (greater than) button.

14. Under **Values**, click **0** (the first value in the list).

15. Click **Execute**.

You selected all the tornadoes that had more than 0 (zero) injuries.

Q26 How many tornadoes caused injuries? (Hint: Look at the Query Results at the bottom of the window.) 83

16. Click the **Statistics** button.

17. Under **Select a field**, scroll down and click **INJ**.

18. Click the box next to **Use Query Results?**

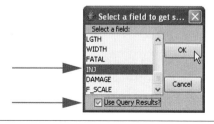

19. Click **OK**. A window called "Statistics Results for field:INJ" opens.

20. Look at the **Statistics Results** window.

21. Look at the bottom row that says **Total**.

 Q27 What is the total number of injuries that occurred? Write this number on your "Historical tornadoes" worksheet.

22. Close the **Statistics Result**s window.

23. In the **Query Builder** window, click **Clear**.

    Now you will find out the total number of deaths that occurred.

24. Under **Select a field**, click **FATAL**.

25. Click the > (greater than) button.

26. Under **Values**, click **0** (the first value in the list).

27. Click **Execute**.

    You selected all the tornadoes that had more than 0 (zero) deaths.

 Q28 How many tornadoes caused deaths? (Hint: Look at the Query Results at the bottom of the window.) 48

28. Click the **Statistics** button.

29. Under **Select a field**, click **FATAL**.

30. Click the box next to **Use Query Results?**

31. Click **OK**. A window called "Statistics Results for field: FATAL" opens.

233

32. Look at the **Statistics Results** window.

33. Look at the bottom row that says **Total**.

Q29 What is the total number of deaths that occurred? Write this number on your "Historical tornadoes" worksheet.

34. Close the **Statistics** window.

35. Close the **Query Builder**.

 36. Click the **Clear All Selections** button.

Now you have filled in the missing information for Event 1.

Q30 Look at your map, and look at your "Historical tornadoes" worksheet. Why do you think the tornado outbreak of April 3–4, 1974, is considered the most violent tornado outbreak in history? List two reasons. Answers may include the following:

- The large number of tornadoes (148)

- The large area affected (13 states and 3 regions)

- The large number of strong (67) and violent (30) tornadoes

- The large number of injuries (5,454) and deaths (310)

Notice that the total damage for this tornado outbreak was more than 600 million dollars. This is the amount it costs to launch the space shuttle!

Step 8: The most costly tornado of all time occurred on May 3, 1999.

1. Turn off the **April 3–4, 1974** layer.

2. Turn on the **May 3, 1999** layer.

Your map shows 71 tornadoes that took place on that day.

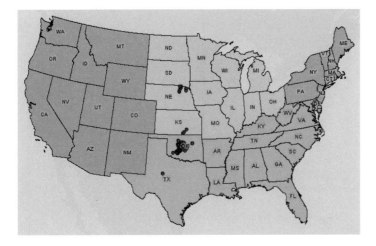

3. Turn on the **Tornado Cities 2** layer. You see the city of Moore, a suburb of Oklahoma City, Oklahoma. This is where the worst tornado took place.

4. Right-click the **Tornado Cities 2** layer name and choose **Attribute Table**. There is one record (row) in the table.

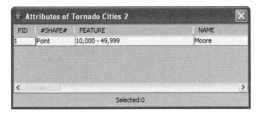

5. Scroll all the way to the right until you see the **HSE_UNITS** field. This field contains the number of housing units (houses).

Q31 How many housing units (houses) are in Moore? **15,801**

6. Close the **Attribute Table**.

The tornado that hit Moore destroyed more than 3,000 of these houses and caused one billion dollars in damage.

7. Right-click the **May 3, 1999** layer name and choose **Attribute Table**.

8. Scroll to the **F_SCALE** field.

9. Right-click **F_SCALE** and choose **Sort Descending**. This reorders the tornadoes from the highest F-scale to the lowest F-scale.

Q32 What is the F-scale of the strongest tornado on this day? F5

10. Click the first record to highlight it. It is also highlighted on the map.

11. Look at the map on your screen.

> **Note:** If the attribute table is covering your map, place your mouse on the words "Attributes of May 3, 1999" at the top of the window. Hold down your mouse button and drag the window out of the way.

You see a yellow dot representing the F5 tornado very close to a white square representing the suburb of Moore. This F5 tornado hit the suburb of Moore on May 3, 1999.

While moving through the area, this F5 tornado created a visible path called a track.

- You can estimate how far this tornado traveled by measuring the length of its track.
- You can estimate the width of this tornado by measuring the width of its track.

12. In the Attribute Table, scroll back until you see the **LGTH** field.

   The numbers in this field represent tenths of a mile. That means that you need to divide the number in the table by 10 to get the actual length (Example: 146 ÷ 10 = 14.6 miles).

 Q33 **How far did this F5 tornado travel? Divide the answer by 10.**

   Length value in table = 370 ÷ 10 = 37.0 miles

13. In the Attribute Table, right-click **LGTH** and choose **Sort Descending**. This reorders the tornadoes from the longest track to the shortest.

Q34 **Did this F5 (highlighted) tornado have the longest track of any tornado that occurred that day (May 3, 1999)?**

   a. Yes

   b. No

14. In the Attribute Table, find the **WIDTH** field (it is right next to the **LGTH** field).

   The numbers in this field represent tens of feet. That means that you need to multiply the number in the table by 10 to get the actual length (Example: 528 x 10 = 5,280 feet or 1 mile). Tornadoes can be more than a mile wide.

 Q35 **What was the width of the F5 tornado? Multiply the answer by 10. (You can ignore the number after the decimal.)**

   Width value in table = 429 X 10 = 4,290 feet

15. In the Attribute Table, right-click **WIDTH** and choose **Sort Descending**. This reorders the tornadoes from the widest to the narrowest.

Q36 **Was this F5 (highlighted) tornado the widest tornado that occurred that day?**

   a. Yes

   b. No

Module 4: Lesson 2

237

Our World GIS Education: Thinking Spatially Using GIS

 How many tornadoes were wider? One

16. Close the **Attribute Table**.

17. Click the **May 3, 1999** layer name to highlight it. It becomes active.

18. Click the **Clear All Selections** button.

19. Look at your "Historical tornadoes" worksheet.

20. Look under the column called **EVENT2**.

   You see many details of the May 3, 1999, tornado outbreak.

   Notice that the number of states, the names of states, and the regions affected are not filled in. You will use the GIS to get this information.

21. Turn off the **Tornado Cities 2** layer.

22. Click the **Select Features** tool and choose **Rectangle**.

   You will use this tool to select all the tornadoes.

23. Click anywhere in the state of Montana (MT) and *hold* down the mouse button.

24. Drag the mouse down and to the right until you reach the state of Louisiana (LA).

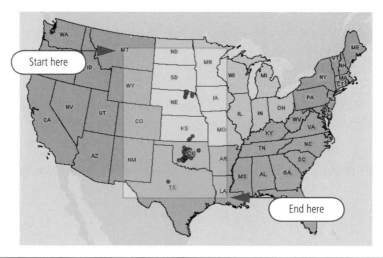

Thinking Spatially Using GIS
Module 4: Lesson 2
18

Module 4: Lesson 2

25. Let go of the mouse button.

    All the tornadoes in the May 3, 1999 layer should be highlighted yellow.

    > **Note:** If you make a mistake, click the **Clear All Selections** button and try again.

    Now you will select all the states that these tornadoes are in.

 26. Click the **Buffer** tool. The Buffer tool creates a one-mile shape around each selected tornado. You will use these buffers to select the states in the layer beneath the tornadoes.

27. Click the box next to **Use buffer to select features from this layer**. A list appears below the box.

28. Select **States** from the list.

29. Click **OK**.

    The GIS uses the buffer (shape) around each tornado to select the state that it is in. The selected states are highlighted in yellow on the map.

30. Click the **States** layer name to highlight it (make it active).

31. Click the **Attributes** button. The Attribute Table for States opens.

32. Look at the bottom of the table next to **Selected**.

    (Q38) **How many states were hit by tornadoes on May 3, 1999? Write your answer on your "Historical tornadoes" worksheet.**

33. In the table, right-click the **STATE_NAME** field and choose **Sort Selected Data to Top**.

239

Q39 Which states were hit by tornadoes on May 3, 1999? Write the names of the states on your "Historical tornadoes" worksheet.

34. Close the **Attribute Table**.

Q40 Look at the legend for the States layer. The highlighted states are located in which regions? Write your answer on your "Historical tornadoes" worksheet.

 35. Click the **Clear All Selections** button.

Now you have filled in the missing information for Event 2.

You can see on your worksheet that the total damage for the May 3, 1999, tornado outbreak was 1.6 billion dollars. The single F5 tornado that hit the Oklahoma City suburb of Moore caused one billion dollars in damage all by itself—it is the most expensive tornado of all time.

Step 9: More tornadoes occurred between May 1 and May 10, 2003, than any other ten-day period since record keeping began.

1. Turn off the **May 3, 1999** layer.

2. Turn on the **May 1-10, 2003** layer.

Your map shows nearly 400 tornadoes that hit the central and southern United States from May 1-10, 2003.

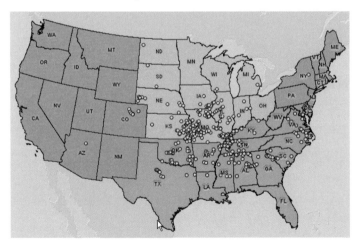

Q41 **Look at your map. Describe in your own words how this tornado outbreak looks different from the other two outbreaks you have already looked at. (Hint: Turn the layers on and off to compare them. How far do the tornadoes extend in all directions? Which regions and states were hit?)** This tornado outbreak is more widespread. It covers a bigger area (more states). It stretches from Arizona in the West to New York and North Carolina in the East.

3. Take out your "Historical tornadoes" worksheet.

4. Look at the column called **EVENT 3**.

You see information about the tornado outbreak of May 1 to May 10, 2003.

On your worksheet, the F-scale and total number of strong and violent tornadoes information is missing. You will use the GIS to get this information.

5. Click the **May 1-10, 2003** layer name to highlight it. It becomes active.

6. Click the **Query Builder** button.

7. Under **Select a field**, choose **F_SCALE**.

8. Click the = (equals) button.

9. Under **Values**, click **2**.

10. Click **Execute**.

11. Look at the **Query Results** at the bottom of the window.

Q42 **How many F2 tornadoes occurred between May 1 and May 10, 2003? Write the number on your "Historical tornadoes" worksheet.**

Now you will follow the same steps to find out the number of F3 tornadoes.

241

12. In the equation box, delete the number **2** and type **3**.

13. Click **Execute**.

14. Look at the **Query Results** at the bottom of the window.

 Q43 How many F3 tornadoes occurred between May 1 and May 10, 2003? Write the number on your "Historical tornadoes" worksheet.

You will follow the same steps to find out the number of F4 tornadoes.

15. In the equation box, delete the number **3** and type **4**.

16. Click **Execute**.

17. Look at the **Query Results** at the bottom of the window.

Q44 How many F4 tornadoes occurred from May 1-10, 2003? Write the number on your "Historical tornadoes" worksheet.

18. There were no F5 tornadoes that took place, so write "**0**" on your handout.

Now that you know the number of F2, F3, F4, and F5 tornadoes, you can use simple addition to find out the number of strong and violent tornadoes. You learned that strong tornadoes have an F-scale of 2 or 3, and violent tornadoes have an F-scale of 4 or 5.

Q45 Add the number of F2 and F3 tornadoes together.

___ ___ (number of F2s) + ___ ___ (number of F3s) = ___ ___ Strong tornadoes. Write this number on your "Historical tornadoes" worksheet.

Q46 Add the number of F4 and F5 tornadoes together.

___ ___ (number of F4s) + ___ ___ (number of F5s) = ___ ___ Violent tornadoes. Write this number on your "Historical tornadoes" worksheet.

19. In the **Query Builder** window, click **Clear**.

20. Close the **Query Builder**.

You have filled in the missing information for Event 3.

Your worksheet shows that the total damage for this tornado outbreak was 3.4 billion dollars, more than the previous two events combined.

Q47 Look at your "Historical tornadoes" worksheet. Why do you think the 2003 tornado outbreak caused more damage than the other two outbreaks? Discuss this with your classmates, and write your answer here. The number of tornadoes from May 1-10 (more than 400 reported) far exceeded the other outbreaks. More cities were involved and more structures sustained damage.

Step 10: Save your work and exit AEJEE.

1. Ask your teacher where and how to save your work.

2. Click the **File menu** and choose **Exit**.

Conclusion

Tornadoes and other natural disasters are events that affect many people in the United States. People are working together to create more effective warning and tracking systems for these disasters in hopes of saving lives.

GIS is one of many tools that help researchers and scientists make educated and effective decisions on predicting when and where these weather patterns will form as well as assist in the recovery process after a disaster has struck.

Module 4, Lesson 2

Analyzing historical tornadoes

Worksheet: Historical tornadoes

	Event 1	Event 2	Event 3
Date	April 3–4, 1974	May 3, 1999	May 1–10, 2003
Why is it remembered?	Largest outbreak	Most expensive single tornado	More tornadoes than any other 10-day period
Total number of tornadoes	148 tornadoes	71 tornadoes	401–412 tornadoes
Cities most affected	Xenia, Ohio	Moore, Oklahoma, (suburb of Oklahoma City)	Kansas City, Kansas; Oklahoma City, Oklahoma; Pierce City, Missouri; Jackson, Tennessee
Total number of states	13 states	**5** states	26 states
Names of states	Alabama, Georgia, Illinois, Indiana, Kentucky, Michigan, Mississippi, New York, North Carolina, Ohio, Tennessee, Virginia, West Virginia	**South Dakota** **Nebraska** **Kansas** **Oklahoma** **Texas**	Alabama, Arizona, Arkansas, Colorado, Georgia, Illinois, Indiana, Iowa, Kansas, Kentucky, Louisiana, Maryland, Michigan, Mississippi, Missouri, Nebraska, New York, North Carolina, North Dakota, Oklahoma, South Carolina, South Dakota, Tennessee, Texas, Virginia, Wisconsin
Regions	Midwest, Northeast, South	**Midwest** **South**	Midwest, Northeast, South, West

Continued on the next page.

	Event 1	Event 2	Event 3
Date	April 3–4, 1974	May 3, 1999	May 1–10, 2003
F-scale	F2 = 34 F3 = 33 F4 = 23 F5 = 7	F2 = 7 F3 = 6 F4 = 3 F5 = 1	F2 = **37** F3 = **17** F4 = **6** F5 = **0**
Total number of strong and violent tornadoes	Strong = 67 Violent = 30	Strong = 13 Violent = 4	Strong = **54** Violent = **6**
Total number of injuries	**5,454** injuries	825 injuries	637 injuries
Total number of deaths	**310** deaths	46 deaths	40+ deaths
Total damage in dollars	>600 million	1.6 billion	3.4 billion

Module 4, Lesson 2

Analyzing historical tornadoes

Handout: Tornado F-Scale

Tetsuya T. Fujita, also known as "Mr. Tornado," invented the Fujita Scale (F-scale) at the University of Chicago in the early 1970s. The F-scale describes different levels of damage caused by tornadoes based on their estimated wind speeds. The scale goes from light damage (F-0) to incredible damage (F-5).

F-scale	Wind speed (miles per hour)	Damage
F0	40–72 mph	**Light damage**—Some damage to chimneys; breaks twigs and tree branches; pushes over some trees; damages signboards; breaks some windows
F1	73–112 mph	**Moderate damage**—Peels surfaces off roofs; mobile homes pushed off foundations; moving autos pushed off the roads; trees snapped or broken
F2	113–157 mph	**Considerable damage**—Roofs torn off houses; mobile homes destroyed; some houses lifted and moved; railroad cars pushed over; large trees snapped or pulled out; light objects take flight
F3	158–206 mph	**Severe damage**—Roofs and walls torn off houses; whole trains turned over; most forest trees pulled out; heavy cars lifted off the ground and thrown; pavement blown off roads
F4	207–260 mph	**Devastating damage**—Houses flattened; weak buildings blown some distance; cars thrown and destroyed; large objects take flight; forest trees pulled out and blown away
F5	261–318 mph	**Incredible damage**—Strong houses lifted and blown away; objects the size of autos fly through the air; bark blown off trees; incredible things happen

Data sources

Module 1 - \OurWorld1\Module 1\data sources include:

\Cntry07 shapefile, from ESRI Data & Maps 2007, courtesy of ArcWorld Supplement.

\continent shapefile, from ESRI Data & Maps 2006, courtesy of ArcWorld Supplement.

\geogrid shapefile, from ESRI Data & Maps 2006, courtesy of ESRI.

\magellan2 shapefile courtesy of the authors.

\minus45plus135 shapefile, from ESRI Data & Maps 2006, courtesy of ESRI.

\Moluccas shapefile, from ESRI Data & Maps 2007, courtesy of ArcWorld Supplement.

\ocean_names shapefile, courtesy of the authors.

\ocean_names2 shapefile, courtesy of the authors.

\Rivers shapefile, from ESRI Data & Maps 2006, courtesy of ArcWorld.

\seville shapefile, from ESRI Data & Maps 2006, courtesy of ArcWorld.

\Stops2 shapefile, courtesy of the authors.

\StudyArea1 shapefile, courtesy of the authors.

\StudyArea2b shapefile, courtesy of the authors.

\StudyArea2c shapefile, courtesy of the authors.

\StudyArea3 shapefile, courtesy of the authors.

\StudyArea_df shapefile, courtesy of the authors.

\world30 shapefile, from ESRI Data & Maps 2006, courtesy of ESRI.

Module 2 - \OurWorld1\Module 1\Data\Pictures sources include:

\Elephant.jpg, courtesy of PhotoDisc/PhotoDisc- Nature, Wildlife and the Environment 2.

\Flamingo.jpg, courtesy of PhotoDisc/PhotoDisc- Nature, Wildlife and the Environment 2.

\Giraffe.jpg, courtesy of PhotoDisc/PhotoDisc- Nature, Wildlife and the Environment 2.

\Gorilla.jpg, courtesy of PhotoDisc/PhotoDisc- Nature, Wildlife and the Environment 2.

\Hippopotamus.jpg, courtesy of PhotoDisc/PhotoDisc- Nature, Wildlife, and the Environment 2.

\Lion.jpg, courtesy of PhotoDisc/PhotoDisc- Nature, Wildlife and the Environment 2.

\Panda.jpg, courtesy of PhotoDisc/PhotoDisc- Nature, Wildlife and the Environment 1.

\Tiger.jpg, courtesy of PhotoDisc/PhotoDisc- Nature, Wildlife and the Environment 2.

\Zebra.jpg, courtesy of PhotoDisc/PhotoDisc- Nature, Wildlife and the Environment 1.

Module 2 - \OurWorld1\Module 2\data sources include:

\AnimalRanges shapefile, courtesy of ArcAtlas.

\Biomes shapefile, courtesy of World Wildlife Fund.

\Boudary shapefile, courtesy of authors.

\Buildings shapefile, courtesy of authors.

\Cntry07 shapefile, from ESRI Data & Maps 2007, courtesy of ArcWorld Supplement.

\Continent shapefile, from ESRI Data & Maps 2006, courtesy of ArcWorld Supplement.

\Geogrid shapefile, from ESRI Data & Maps 2006, courtesy of ESRI.

\Habitats shapefile, courtesy of authors.

\Status shapefile, courtesy of World Wildlife Fund.

\Walkway shapefile, courtesy of authors.

\World30 shapefile, from ESRI Data & Maps 2006, courtesy of ESRI.

\ZooAnimals shapefile, courtesy of the authors.

Module 3 - \OurWorld1\Module 3\data sources include:

\cities shapefile, from ESRI Data & Maps 2006, courtesy of US Census.

\comp_usage shapefile, courtesy of US Census.

\lakes shapefile, from ESRI Data & Maps 2006, courtesy of ArcWorld.

\NERivers shapefile, from ESRI Data & Maps 2006, courtesy of ArcWorld.

\poly5000 shapefile, from AEJEE, courtesy of USGS and Digital Chart of the World.

\prism0p020_Dis2 shapefile, courtesy of National Atlas.

\Rivers shapefile, from ESRI Data & Maps 2006, courtesy of ArcWorld.

\sm_cntry shapefile, from ESRI Data & Maps 2006.

\StateLevel1 shapefile, from ESRI Data & Maps 2006, courtesy of ArcUSA, US Census, and ESRI.

\StateLevel3 shapefile, from ESRI Data & Maps 2006, courtesy of ArcUSA, US Census, and ESRI.

\states shapefile, from ESRI Data & Maps 2006, courtesy of ArcUSA, US Census, and ESRI.

\states_1 shapefile, from ESRI Data & Maps 2006, courtesy of ArcUSA, US Census, and ESRI.

\US48SHD2.jpg, from AEJEE, courtesy of USGS.

\US48SHD2.prj, from AEJEE, courtesy of USGS.

\world30 shapefile, from ESRI Data & Maps 2006, courtesy of ESRI.

\WWF_terra_Clip shapefile, courtesy of World Wildlife Fund.

Module 4 - \OurWorld1\Module 4\data sources include:

\1974_2003_T shapefile, courtesy of National Atlas.

\April3_4_74 shapefile, courtesy of National Atlas.

\Density_states2, courtesy of National Atlas.

\F3to5_95to04 shapefile, courtesy of National Atlas.

\Fall_T shapefile, courtesy of National Atlas.

\lakes shapefile, from ESRI Data & Maps 2006, courtesy of ArcWorld.

\May1_10_2003 shapefile, courtesy of National Atlas.

\May3_1999 shapefile, courtesy of National Atlas.

\Moore shapefile, data from ESRI Data & Maps 2006, courtesy of US Census, combined with data courtesy of National Atlas.

\Names2 shapefile, courtesy of the authors.

\sm_cntry shapefile, from ESRI Data & Maps 2006.

\Spring_T shapefile, courtesy of National Atlas.

\State_tornadoes shapefile, from ESRI Data & Maps 2006, courtesy of ArcUSA, US Census, and ESRI.

\states_1 shapefile, from ESRI Data & Maps 2006, courtesy of ArcUSA, US Census, and ESRI.

\states_1_Dissolve, from ESRI Data & Maps 2006, courtesy of ArcUSA, US Census, and ESRI.

\Strong_T40 shapefile, courtesy of National Atlas.

\Summer_T shapefile, courtesy of National Atlas.

\Torn9504states, courtesy of National Atlas.

\torn_cities3 shapefile, data from ESRI Data & Maps 2006, courtesy of US Census, combined with data courtesy of National Atlas.

\US48SHD2.jpg, from AEJEE, courtesy of USGS.

\US48SHD2.prj, from AEJEE, courtesy of USGS.

\Violent_T40 shapefile, courtesy of National Atlas.

\Weak2_40 shapefile, courtesy of National Atlas.

\Winter_T shapefile, courtesy of National Atlas.

\world30 shapefile, from ESRI Data & Maps 2006, courtesy of ESRI.

\Xenia shapefile, data from ESRI Data & Maps 2006, courtesy of US Census, combined with data courtesy of National Atlas.

Image credits

Introduction
Page xii: Courtesy of Comstock/Comstock-Kids/Jupiterimages

Module 1
Page 12: Courtesy of PhotoDisc/PhotoDisc-Maps and Navigation

Module 2
Page 62: Courtesy of PhotoDisc/PhotoDisc-Nature, Wildlife and the Environment 2

Module 3
Page 128: Courtesy of Comstock/Comstock-Wild West/Jupiterimages

Module 4
Page 182: Courtesy of NOAA Photo Library, NOAA Central Library; OAR/ERL/National Severe Storms Laboratory (NSSL)

Data license agreement

Important: Read carefully before opening the sealed media package

Environmental Systems Research Institute, Inc. (ESRI), is willing to license the enclosed data and related materials to you only upon the condition that you accept all of the terms and conditions contained in this license agreement. Please read the terms and conditions carefully before opening the sealed media package. By opening the sealed media package, you are indicating your acceptance of the ESRI License Agreement. If you do not agree to the terms and conditions as stated, then ESRI is unwilling to license the data and related materials to you. In such event, you should return the media package with the seal unbroken and all other components to ESRI.

ESRI License Agreement

This is a license agreement, and not an agreement for sale, between you (Licensee) and Environmental Systems Research Institute, Inc. (ESRI). This ESRI License Agreement (Agreement) gives Licensee certain limited rights to use the data and related materials (Data and Related Materials). All rights not specifically granted in this Agreement are reserved to ESRI and its Licensors.

Reservation of Ownership and Grant of License: ESRI and its Licensors retain exclusive rights, title, and ownership to the copy of the Data and Related Materials licensed under this Agreement and, hereby, grant to Licensee a personal, nonexclusive, nontransferable, royalty-free, worldwide license to use the Data and Related Materials based on the terms and conditions of this Agreement. Licensee agrees to use reasonable effort to protect the Data and Related Materials from unauthorized use, reproduction, distribution, or publication.

Proprietary Rights and Copyright: Licensee acknowledges that the Data and Related Materials are proprietary and confidential property of ESRI and its Licensors and are protected by United States copyright laws and applicable international copyright treaties and/or conventions.

Permitted Uses: Licensee may install the Data and Related Materials onto permanent storage device(s) for Licensee's own internal use.

Licensee may make only one (1) copy of the original Data and Related Materials for archival purposes during the term of this Agreement unless the right to make additional copies is granted to Licensee in writing by ESRI.

Licensee may internally use the Data and Related Materials provided by ESRI for the stated purpose of GIS training and education.

Uses Not Permitted: Licensee shall not sell, rent, lease, sublicense, lend, assign, time-share, or transfer, in whole or in part, or provide unlicensed Third Parties access to the Data and Related Materials or portions of the Data and Related Materials, any updates, or Licensee's rights under this Agreement.

Licensee shall not remove or obscure any copyright or trademark notices of ESRI or its Licensors.

Term and Termination: The license granted to Licensee by this Agreement shall commence upon the acceptance of this Agreement and shall continue until such time that Licensee elects in writing to discontinue use of the Data or Related Materials and terminates this Agreement. The Agreement shall automatically terminate without notice if Licensee fails to comply with any provision of this Agreement. Licensee shall then return to ESRI the Data and Related Materials. The parties hereby agree that all provisions that operate to protect the rights of ESRI and its Licensors shall remain in force should breach occur.

Disclaimer of Warranty: The Data and Related Materials contained herein are provided "as-is," without warranty of any kind, either express or implied, including, but not limited to, the implied warranties of merchantability, fitness for a particular purpose, or noninfringement. ESRI does not warrant that the Data and Related Materials will meet Licensee's needs or expectations, that the use of the Data and Related Materials will be uninterrupted, or that all nonconformities, defects, or errors can or will be corrected. ESRI is not inviting reliance on the Data or Related Materials for commercial planning or analysis purposes, and Licensee should always check actual data.

Data Disclaimer: The Data used herein has been derived from actual spatial or tabular information. In some cases, ESRI has manipulated and applied certain assumptions, analyses, and opinions to the Data solely for educational training purposes. Assumptions, analyses, opinions applied, and actual outcomes may vary. Again, ESRI is not inviting reliance on this Data, and the Licensee should always verify actual Data and exercise their own professional judgment when interpreting any outcomes.

Limitation of Liability: ESRI shall not be liable for direct, indirect, special, incidental, or consequential damages related to Licensee's use of the Data and Related Materials, even if ESRI is advised of the possibility of such damage.

No Implied Waivers: No failure or delay by ESRI or its Licensors in enforcing any right or remedy under this Agreement shall be construed as a waiver of any future or other exercise of such right or remedy by ESRI or its Licensors.

Order for Precedence: Any conflict between the terms of this Agreement and any FAR, DFAR, purchase order, or other terms shall be resolved in favor of the terms expressed in this Agreement, subject to the government's minimum rights unless agreed otherwise.

Export Regulation: Licensee acknowledges that this Agreement and the performance thereof are subject to compliance with any and all applicable United States laws, regulations, or orders relating to the export of data thereto. Licensee agrees to comply with all laws, regulations, and orders of the United States in regard to any export of such technical data.

Severability: If any provision(s) of this Agreement shall be held to be invalid, illegal, or unenforceable by a court or other tribunal of competent jurisdiction, the validity, legality, and enforceability of the remaining provisions shall not in any way be affected or impaired thereby.

Governing Law: This Agreement, entered into in the County of San Bernardino, shall be construed and enforced in accordance with and be governed by the laws of the United States of America and the State of California without reference to conflict of laws principles. The parties hereby consent to the personal jurisdiction of the courts of this county and waive their rights to change venue.

Entire Agreement: The parties agree that this Agreement constitutes the sole and entire agreement of the parties as to the matter set forth herein and supersedes any previous agreements, understandings, and arrangements between the parties relating hereto.

Installing the software and data on Microsoft Windows

Thinking Spatially Using GIS includes one CD at the back of the book. This CD contains the GIS software, lesson data, and teacher resources you will need.

Installation of the software and data on Windows requires approximately 215 megabytes of disk space. The software requires Microsoft Windows XP or Windows 2000. It does not run on Windows Vista. At least 256 MB RAM is required.

Installation of the teacher resources requires approximately 10 megabytes of additional disk space. Many of the teacher resource documents are in PDF format. To read these you will need Adobe Reader 6.0 or higher.

Installing the AEJEE software on Windows

This book comes with ArcExplorer Java Edition for Education (AEJEE) version 2.3.2. If you already have AEJEE 2.3.2 installed on your computer, you can skip this step and go to "Installing the lesson data." If you have an older version, you must uninstall the older version first. (If you aren't sure how to uninstall the software, turn to "Uninstalling the software, data, and teacher resources" on page 261)

Follow the steps below to install the software:

1. Put the CD in your computer's CD drive. A window like the one below will appear.

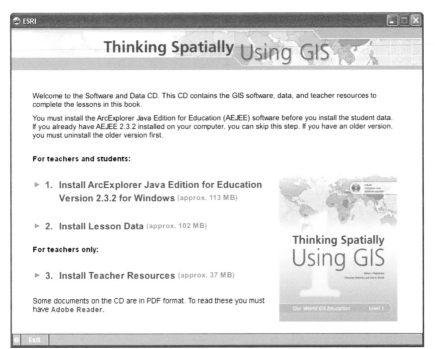

2. Read the welcome, and then click Install ArcExplorer Java Edition for Education version 2.3.2 (Windows).

3. Read the information on the next panel, and then click the link. This launches the InstallAnywhere program. After preparing to install, the ArcExplorer Java Edition for Education Introduction screen appears.

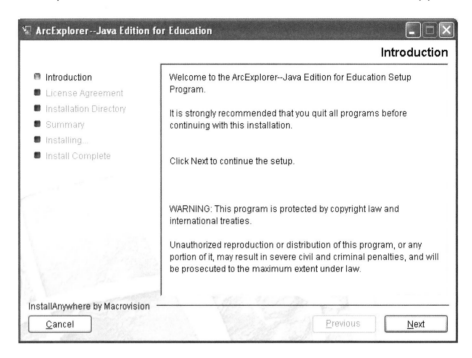

4. Click Next. Read and accept the license agreement terms.

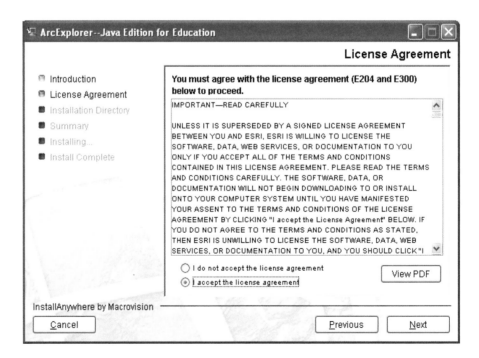

5. Click Next. You will accept the default installation folder (**C:\ESRI**).

6. Click Next. The Pre-Installation Summary screen displays.

7. Click Install. The installation will take a few moments. When the installation is complete, you will see the following message:

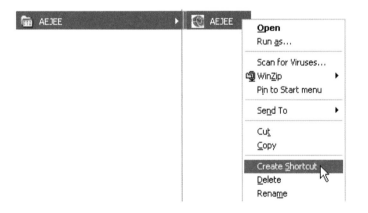

8. Click Done. The software is installed on your computer in **C:\ESRI\AEJEE**.

9. To start the software, click Start > All Programs > AEJEE > AEJEE.

10. To create an AEJEE icon on the desktop, go to Start > All Programs > AEJEE. Right-click AEJEE and choose Create Shortcut. Drag the shortcut item to the desired location on the desktop.

Installing the lesson data

Follow the steps below to install the folder:

1. Click Home at the top of the Welcome window if it is still open. Otherwise, put the CD in your computer's CD drive. A window like the one below will appear.

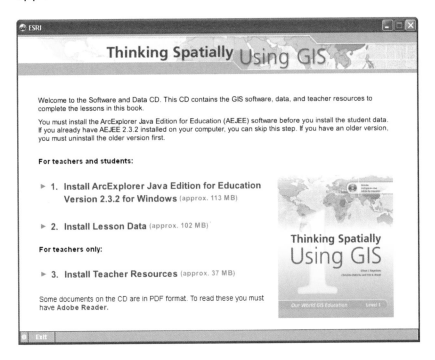

2. Read the welcome, and then click Install lesson data.

3. Read the information on the next panel, and then click the link. This launches the InstallShield Wizard.

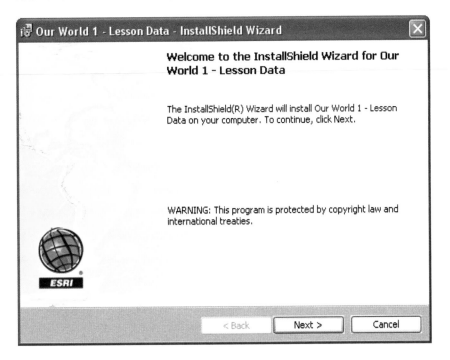

4. Click Next. Read and accept the license agreement terms.

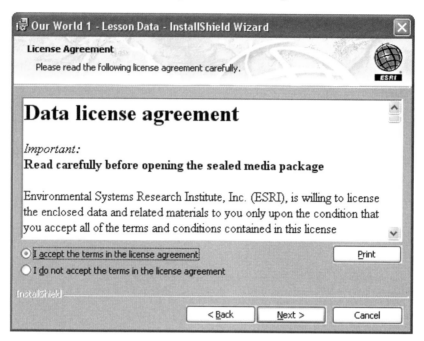

5. Click Next. You will accept the default installation folder.

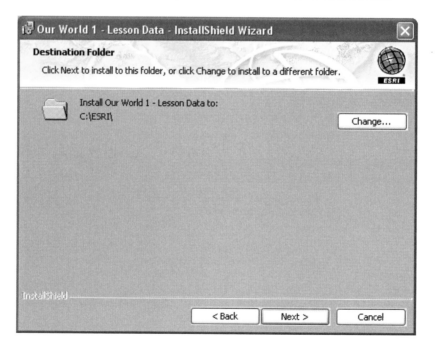

6. Click Next. The installation will take a few moments. When the installation is complete, you will see the following message:

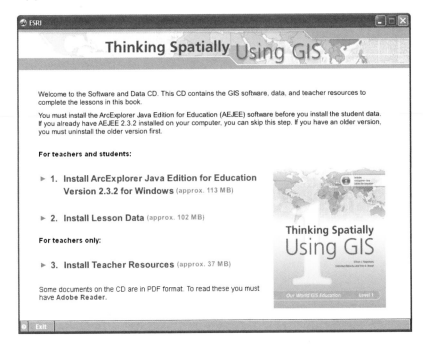

7. Click Finish. The lesson data is installed in **C:\ESRI\AEJEE\Data\OurWorld1**.

Installing the teacher resources

Follow the steps below to install the folder:

1. Click Home at the top of the Welcome window if it is still open. Otherwise, put the CD in your computer's CD drive. A window like the one below will appear.

2. Read the welcome, and then click Install teacher resources.

3. Read the information on the next panel, and then click the link. This launches the InstallShield Wizard.

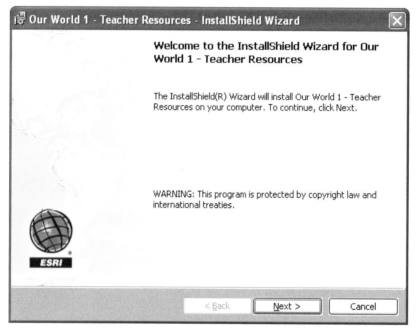

4. Click Next. Accept the default installation folder (**C:\ESRIPress**) or click Change and navigate to the drive or folder location where you want to install the teacher resources.

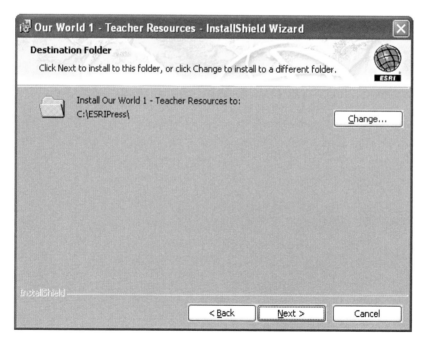

5. Click Next. The installation will take a few moments. When the installation is complete, you will see the following message:

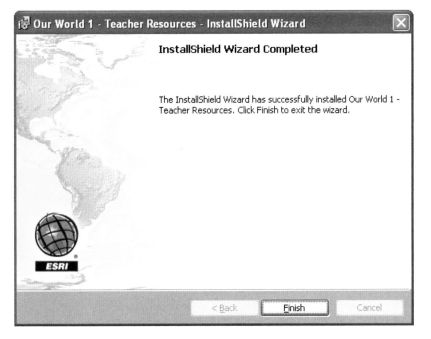

6. Click Finish. The teacher resources are installed in a folder called **OurWorld1_teacher**.

Your installation is complete.

Uninstalling the software, data, and teacher resources

To uninstall the AEJEE software (and the lesson data, if installed), open the operating system's control panel and double-click the Add/Remove Programs icon. In the Add/Remove Programs dialog box, select the following entry and follow the prompts to remove it:

ArcExplorer Java Edition for Education

To uninstall the lesson data only, open the operating system's control panel and double-click the Add/Remove Programs icon. In the Add/Remove Programs dialog box, select the following entry and follow the prompts to remove it:

Our World 1—Lesson Data

To uninstall the teacher resources, open the operating system's control panel and double-click the Add/Remove Programs icon. In the Add/Remove Programs dialog box, select the following entry and follow the prompts to remove it:

Our World 1—Teacher Resources

Installing the software and data on Macintosh

Thinking Spatially Using GIS includes one CD at the back of the book. This CD contains the GIS software, lesson data, and teacher resources you will need.

Installation of the software and data on Macintosh requires approximately 102 megabytes of disk space. The software requires Macintosh OS 10.2 minimum and at least 256 MB RAM.

Installation of the teacher resources requires approximately 37 megabytes of additional disk space. Many of the teacher resource documents are in PDF format. To read these you will need Adobe Reader 6.0 or higher.

We recommend that you begin by reading the Readme_OWInstall_for_Mac.txt file on the CD:

1. Put the CD in your computer's CD drive. Double-click the CD icon that appears on your screen to open the Finder.

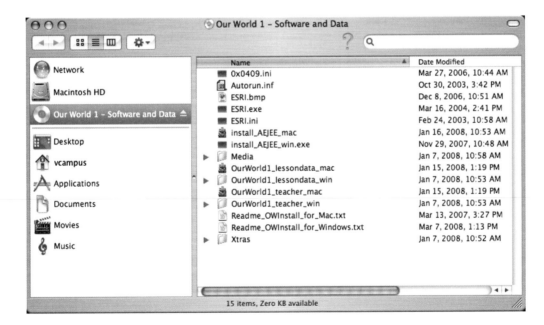

2. Double-click the Readme_OWInstall_for_Mac.txt file and read it. Close the file when finished.

Installing the AEJEE software on Macintosh

This book comes with ArcExplorer Java Edition for Education (AEJEE) version 2.3.2. If you already have AEJEE 2.3.2 installed on your computer, you can skip this step and go to "Installing the lesson data." If you have an older version, you must uninstall the older version first. (If you aren't sure how to uninstall the software, turn to "Uninstalling the software, data, and resources" on page 272)

Follow the steps below to install the software:

1. Return to the Finder if it is still open. Otherwise, make sure the CD is in your computer's CD drive. Double-click the CD icon to open the Finder.

2. Double-click the install_AEJEE_mac icon. This launches the InstallAnywhere program. The ArcExplorer Java Edition for Education Introduction screen appears.

3. Click Next. Read and accept the license agreement terms.

4. Click Next. You will accept the default installation folder (**[hard drive]/ESRI**).

5. Click Next. The Pre-Installation Summary screen displays.

6. Click Install. The installation will take a few moments. When it is complete, you will see the following message:

7. Click Done. The software is installed on your computer.

8. To start the software, go to the **[hard drive]/ESRI/AEJEE** folder. Double-click the AEJEE icon.

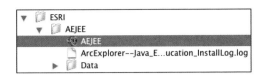

9. To create an AEJEE alias on the desktop, Control-click AEJEE in the Finder and choose Make Alias. Drag the alias icon to the desired location on your desktop or your Dock.

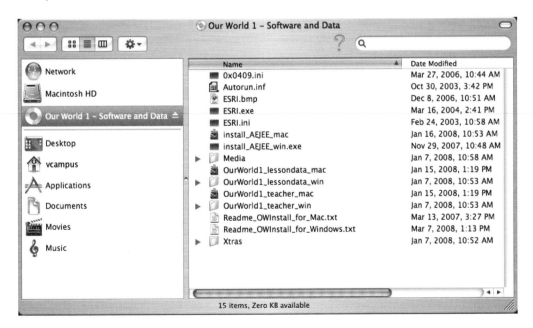

Note: Before starting the lessons, be sure to read the Macintosh User Guide in this book.

Installing the lesson data

Follow the steps below to install the folder:

1. Make sure the CD is in your computer's CD drive. Double-click the CD icon to open the Finder.

2. Double-click the **OurWorld1_lessondata_mac** icon. This launches the InstallAnywhere program. The Our World 1 Welcome screen appears.

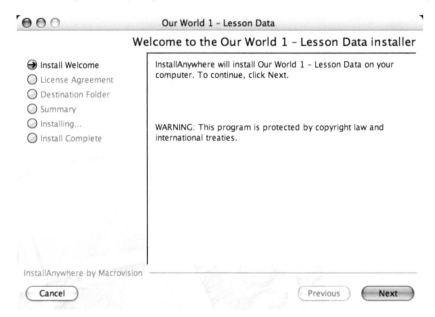

3. Click Next. Read and accept the license agreement terms.

4. Click Next. You will accept the default destination folder.

5. Click Next. The Summary screen displays. Review the summary.

6. Click Install. The installation will take a few moments. When the installation is complete, you will see the following message:

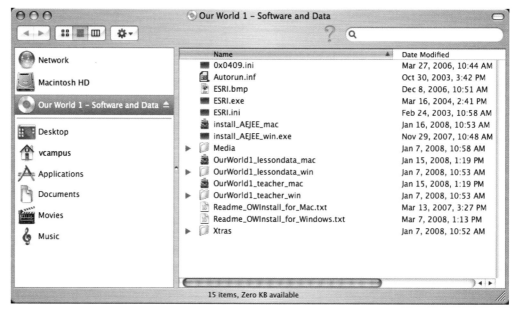

7. Click Done. The lesson data is installed in **[hard drive]/ESRI/AEJEE/Data/ OurWorld1**.

Installing the teacher resources

Follow the steps below to install the folder:

1. Make sure the CD is in your computer's CD drive. Double-click the CD icon to open the Finder.

2. Double-click **OurWorld1_teacher_mac**. This launches the InstallAnywhere program. The Our World 1 – Teacher Resources welcome screen appears.

3. Click Next. Accept the default destination folder or click Choose and navigate to the drive or folder location where you want to install the teacher resources.

4. Click Next. The Summary screen appears. Review the summary.

5. Click Install. The installation will take a few moments. When the installation is complete, you will see the following message:

6. Click Done. The teacher resources are installed in **[hard drive]/ESRIPress/ OurWorld1_teacher**.

Your installation is complete.

Uninstalling the software, data, and resources

To uninstall the AEJEE software, navigate to the [hard drive]/ESRI/AEJEE/Uninstall_ArcExplorer Java Edition for Education. Double-click the Uninstall ArcExplorer Java Edition for Education icon. Follow the prompts in the uninstall wizard to remove the program.

To uninstall the lesson data, navigate to the [hard drive]/ESRI/AEJEE/Data/Uninstall_Our World 1 – Lesson Data folder. Double-click the Uninstall Our World 1 – Lesson Data icon. Follow the prompts in the uninstall wizard to remove the data.

To uninstall the teacher resources, navigate to the [hard drive]/ESRIPress/OurWorld1_teacher/ Uninstall_Our World 1 – Teacher Resources folder. Double-click the Uninstall Our World 1 – Teacher Resources icon. Follow the prompts in the uninstall wizard to remove the data.

Macintosh user guide

In this book, all the illustrations show how AEJEE software looks under the Microsoft Windows operating systems. The appearance and operation of AEJEE is nearly the same on both Windows and Macintosh platforms. However, there are some cosmetic differences and a few procedural differences. This document describes the main differences that Macintosh users will encounter. When a Mac user's screen differs from what is shown in this book or when the instructions for a particular step differ, reference to the information below will clarify and reconcile the difference.

Application window

When you start AEJEE, it will appear with the menu bar across the top of the screen and the map window in the middle. AEJEE has a feature that allows you to change the way the application window looks to mimic the way it looks on a Windows computer.

This procedure must be done before beginning the lesson:

1. Start the AEJEE software.

2. Click the Window menu.

3. Click **Look and Feel**.

4. Choose **Metal**.

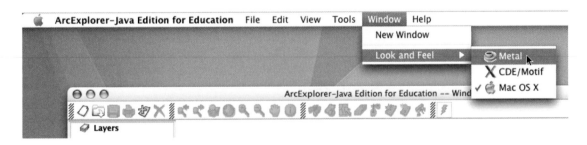

This will automatically change the look of the software and make it appear as though you are working in a Windows environment. It will bring the menu bar together with the map window. Then your screen will closely match the illustrations shown in this book.

Note: You (or your students) will need to perform this procedure each time you start the AEJEE software.

Maximizing and closing windows

On Macintosh, you close a window by clicking the red (x) button in the upper left corner of the window. To Maximize or restore a window, you click the green (+) button. You minimize a window by clicking the yellow (–) button.

To maximize the AEJEE window, click the green (+) button

Opening project files

After changing the look and feel to Metal and maximizing the AEJEE window, you are ready to open a project (.axl) file. The project file contains the map and the data layers needed to complete the lesson. The procedure for opening a project file is the same on a Macintosh or Windows computer, but the Open window looks slightly different on Macintosh (see below).

1. Click the **Open** button on the AEJEE interface. The Open window appears.

2. Choose **OurWorld1** and click **Open**.

3. Choose **Module1** and click **Open**.

4. Choose **MagellanAtlantic.axl** and click **Open**.

A map of the world appears on your screen.

Right-clicking

On a Windows computer, you open a menu for a layer by right-clicking the layer name; to choose an option from the menu, you left-click.

On a Macintosh computer, you can perform these same operations with a one-button mouse. To simulate right-clicking a layer, hold down the Control key on the keyboard while you click the layer name. To choose a menu option, just click the option (without holding down the Control key).

On Windows, you right-click the Stops layer name to display its menu; on Macintosh, you hold down the Control key as you click the layer name.

Selecting multiple records in a table

On a Windows computer, you can select multiple adjacent records in a table by holding down the Ctrl key on your keyboard as you left-click each of the records.

On a Macintosh computer, you can perform the same operation by holding down the Shift key on your keyboard as you click each of the records.

Attributes of % Internet Access

COMP_TOT	INTACC_TOT	INTACC_PCT	COMACC_PC
537	435	39.55	48.82
544	462	42.31	49.82
935	785	44.1	52.53
390	323	44.86	54.17
996	846	45.8	53.93
899	749	45.95	55.15
409	354	47.39	54.75
787	687	48.31	55.34
1357	1171	49.02	56.8
951	813	49.91	58.38

Selected:3

On Windows, you hold down the Ctrl key as you click each record; on Macintosh, you hold down the Shift key as you click each record.

The ArcExplorer™ Java Edition for Education (AEJEE) software accompanying this book is provided for educational purposes. AEJEE software does not run on the Microsoft® Windows® Vista operating system.

The book and media are returnable only if the original seal on the software media packaging is not broken and the media packaging has not otherwise been opened.

> **Please carefully read the instructions on the previous pages for important information about installing the software, data, and teacher resources from the CD. The software must be installed before you install the lesson data.**

If you have problems installing the software, lesson data, or teacher resources, check the FAQ section of this book's Web site, www.esri.com/ourworldgiseducation. Or, send an email with your questions to ESRI workbook support at learngis@esri.com.